AF289861

B.-M. Uebele-Kallhardt

# Human Oocytes and Their Chromosomes

## An Atlas

In Cooperation with T. Trautmann

With a Foreword by K. Benirschke

With 72 Figures

Springer-Verlag
Berlin  Heidelberg  New York  1978

BERTA-MARGARETA UEBELE-KALLHARDT, Dr. rer. nat.
Scientific Collaborator
Department of Obstetrics and Gynecology
Department of Clinical Genetics
University of Ulm, West Germany


THEA TRAUTMANN, Technical Assistent
Cytogenetic Laboratory
Department of Clinical Genetics
University of Ulm, West Germany

ISBN-13: 978-3-642-95330-9     e-ISBN-13: 978-3-642-95328-6
DOI: 10.1007/978-3-642-95328-6

Library of Congress Cataloging in Publication Data. Uebele-Kallhardt, B.-M. 1913- Human oocytes and their chromosomes. Bibliography: p. Includes index. 1. Oogenesis —Atlases. 2. Human chromosomes —Atlases. I. Trautmann, Thea, joint author. II. Title. QL965.U34 612.6′2    78-14350



2121/3130-543210

# Foreword

The last decade has seen remarkable advances in human genetics. Once the correct chromosome number of the human genome was ascertained, a wide variety of diseases was recognized as due to numerical chromosome anomalies. There followed the discovery that spontaneous abortions are the result of chromosome errors, and specific band patterns of chromosomes allowed identification of minute lesions. The techniques of cell hybridization now allow specific gene assignment to chromosomes and even to distinct loci on their arms. All this was possible because of the ease with which metaphase chromosomes can be obtained and manipulated. The much older technique of analysis of meiotic chromosomes has taken a back seat in this exciting era. Being much less readily accessible, spermatogonial analysis is much less frequently undertaken and is less successful. Even more difficult for study is the female meiotic process. Not only is meiosis extraordinarily long, spanning from before birth to ovulation, the techniques for its study and the patience required for detailed inquiry have been significant obstacles. At the same time, the suspicion that female meiotic analysis should not only be rewarding but that it may be mandatory has been with us ever since it was recognized that a positive correlation exists between chromosomal nondisjunction and maternal age.

Before the intricacies of chromosomal behavior that are responsible for nondisjunction are understood, however, it is necessary that we comprehend the normalcy of the process. That is the aim of this presentation. A systematic inquiry of human oogonial maturation has been made in only two or three laboratories. The cumulative descriptive results from one of these patient efforts now lie before us. Here are the detailed photographic records of the maturing oocyte, accompanied by descriptions which reflect the interpretation by the author. Having experience with an unusually large material she describes the initial meiotic prophases in fetuses after eighteen weeks of gestation to birth when the first phase is arrested. The continuation of meiosis is followed in biopsies of adult ovaries from which the oocytes are dissected and allowed to mature in tissue culture explants. A number of normal and abnormal chromosome sets could be identified in this tenacious study and will give us the background for comparison. A comprehensive citation of literature and succinct technical details are woven into the presentation to make the book more useful to future workers.

This is an unusual book. It is not meant to be read as other texts. It will be useful for the laboratory bench of cytogeneticists with an interest in exploring the borderline of knowledge and the wish to extend this imperfect understanding we now have of this important process. It is an atlas of beautiful photographs that sequentially depicts the meiotic progression and brings together all that is currently known of oogenesis. The author is to be congratulated for having persevered in so difficult a task as the accumulation and interpretation of this vast material. We also owe a special thanks to the publishers for presenting a book of this quality to a necessarily limited audience so that science may proceed at a more rapid pace. We wish that the book may find wide acceptance as a milestone in human cytogenetics.

San Diego, California                                    K. Benirschke

# Acknowledgments

The author would like to express her gratitude to Professor K. Knörr, head of the Department of Obstetrics and Gynecology of the University of Ulm. His animating interest in genetic questions made it possible for me to carry out these investigations.

Thanks are due to all the members of this department, especially the surgical staff, for having provided me with the ovarian material.

Finally I am most grateful to Dr. R.G. Edwards, Physiological Laboratory, Cambridge University (U.K.), who, in 1969, introduced me to the intricate art of handling and culturing mammalian oocytes.

B.-M. Uebele-Kallhardt

This study was supported by grants from the Deutsche Forschungsgemeinschaft, Bonn–Bad Godesberg.

# Contents


Introduction . . . . . . . . . . . . . . . . . . . . . . 1

Oocytes of Fetal Ovaries:
Prophase of First Meiotic Division . . . . . . . . . . 3

Selected References . . . . . . . . . . . . . . . . . 37

Oocytes of Adult Ovaries:
First and Second Meiotic Divisions . . . . . . . . . . 41

Selected References . . . . . . . . . . . . . . . . . 95

Materials and Methods . . . . . . . . . . . . . . . 99

Terminology . . . . . . . . . . . . . . . . . . . . 102

Selected References . . . . . . . . . . . . . . . . 103

Subject Index . . . . . . . . . . . . . . . . . . . 105

# Introduction

Meiosis is by far the most essential part of oogenesis. It consists of two cell divisions, known as first and second maturation divisions, during which the nucleus and the cytoplasm undergo a number of changes. The highly specialized meiotic process causes an exchange of genetic material between pairs of chromosomes and provides the mature oocyte with half the number of chromosomes characteristic of the somatic cells.

Meiosis is of exceptionally long duration in the human oocyte. In the fetal ovary, oocytes enter the prophase of the first maturation division. At the time of birth all primary oocytes are arrested in a stage, known as dictyotene, that persists for many years. In the adult ovary this prolonged stage of development is terminated just before ovulation, when an individual oocyte resumes meiosis and begins its final maturation. The second maturation division is completed only if fertilization occurs.

Detailed knowledge of normal meiotic behavior of the chromosomes and its particulars (chromosome pairing, crossing-over, chiasma formation, separation and distribution of homologous chromosomes) is necessary for a full understanding of chromosomal abnormalities and their origins, problems which confront researchers in clinical genetics. As an example, a case of translocation heterozygosity will be treated extensively in this report.

Due to varying degrees of spiralization, the coiling and uncoiling of the chromosome threads, striking phenotypic changes are observable in the chromosomes during meiosis. The maximally despiralized and extended chromosomes represent the functional form and are considered to be suggestive of intense transcriptional activity. The coiling cycle transforms the functional form into the transport form, in which the chromosomes are extremely spiralized and contracted. The preovulatory stages of meiosis are now available for study by in vitro maturation of oocytes. It seems worthwhile to present in its entirety the meiotic coiling cycle of the chromosomes from the prophase of meiosis up to the metaphase of the second maturation division. By means of microphotographs, a comprehensive picture of the sequential nuclear stages and the related chromosomal changes which occur in the human oocyte during meiosis in the fetal and adult ovaries is given.

Meiotic behavior of the human oocyte is generally in accordance with the observed basic rules of meiosis. Usually these rules are illustrated by schematic drawings. However, the simplification and subdivision necessary in such drawings is more or less arbitrary and does not do justice to the complex and dynamic process of meiosis. It is for these reasons, among others, that this report will show more life-like representations of meiosis and of the coiling cycle of the chromosomes within the human oocyte.

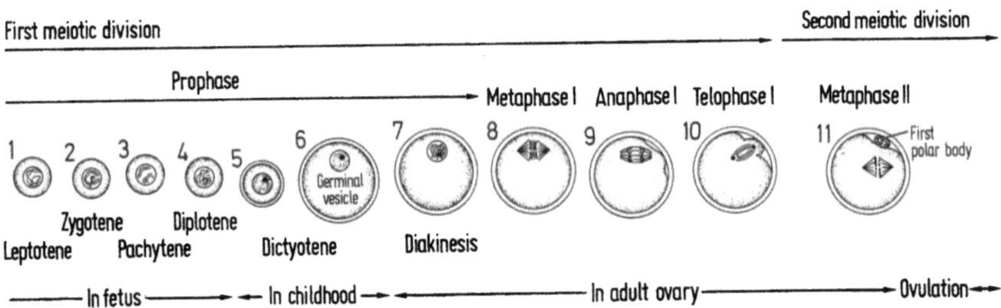


Fig. 1. Diagrammatic representation of the oocyte maturation process from the prenatal period to the preovulatory stage (modified from Edwards, R.G.: Mammalian eggs in the laboratory, Sci. Am. **215**, 73–81 (1966)

The micrographs on the following pages illustrate the sequence of maturational changes that human oocytes undergo during meiosis.

2

# Oocytes of Fetal Ovaries:
# Prophase of First Meiotic Division

Between the 3rd and 7th month in the fetal ovary, there is not only an increase of germ cells by mitotic divisions, but also the transformation of oogonia into oocytes.

Having completed the last premeiotic interphase, during which DNA replication takes place, the germ cells, growing in size, enter the *preleptotene* stage of the first meiotic prophase. As has been shown in recent years, this phase of transition to the leptotene is characterized by profound changes in the chromosome shape, caused by increasing spiralization up to maximum condensation in the so-called prochromosomes and by the despiralization that follows.
The fine filaments of the *leptotene* nucleus emerge from this phase of despiralization. Subsequently the chromosomes once again contract and enter the *zygotene* stage, in which pairing of homologous chromosomes begins. In the next stage, the *pachytene,* the chromosomes pair completely (synapsis). These pairs are called bivalents, their number equal to that of half of the somatic chromosomes. Due to increasing spiralization, the chromosomes become shorter and thicker, and at this point a number of pairs can be identified. This is especially true of the acrocentric bivalents associated with nucleoli. At the end of the relatively long pachytene stage an exchange of genetic material takes place between the paired homologous chromosomes (crossing-over). In the subsequent *diplotene* stage, the oocyte's chromosomes once again despiralize. The homologues repel one another. Only the chiasmata, i.e., the sites of genetic exchange, hold the pairing partners together for the present.

At this stage in prophase, meiosis of the human female germ cell is suspended until sexual maturity and ovulation. The oocyte remains in a modified diplotene stage, the so-called *dictyotene*. The uncoiled, extremely extended bivalents, known as "lampbrush chromosomes," are difficult to observe even with the aid of an electron microscope. Their specific structure as well as the nucleoli, one to four per cell present throughout the entire prophase, are related to the intense metabolic and synthetic activity of the primary oocyte. By accumulation of nutrients, the oocyte increases its cytoplasm and expands greatly in size, finally reaching a diameter of 100–120 μ.

The suspension of meiosis in the dictyotene stage until the preovulatory phase of the oocyte's development probably occurs under the influence of epithelial cells, which together with the oocyte form the primary follicle at the end of the first meiotic prophase.

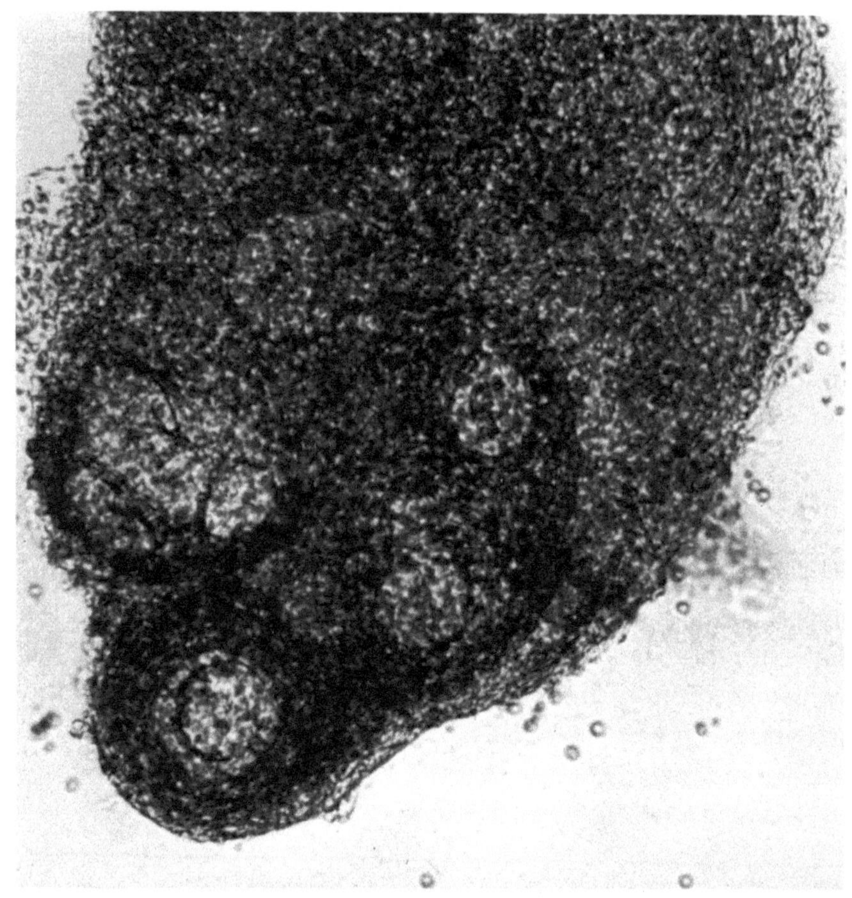

Fig. 2. Segment of an explanted fetal ovary. A group of germ cells, different in size, is present

Unstained, ×410

## Oogonia

Fig. 3. Nucleus of an oogonium with its centrally situated nucleolus (*arrow*)

× 1,600

Oogonia, derived from primordial germ cells, undergo mitotic divisions, their number of chromosomes remaining at 46. At the end of the mitotic proliferation, oogonia become transformed into primary oocytes, when they enter the meiotic prophase.

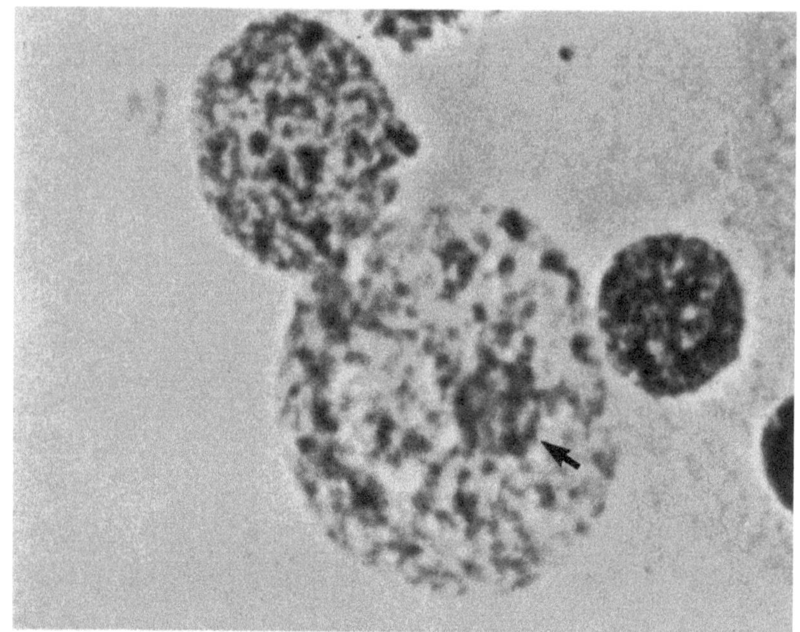

Fig. 3

Fig. 4. Nuclei of primary oocytes in different stages of the first meiotic prophase: preleptotene (prochromosomes), leptotene, pachytene. The smaller-sized nuclei are derived from somatic cells

× 600

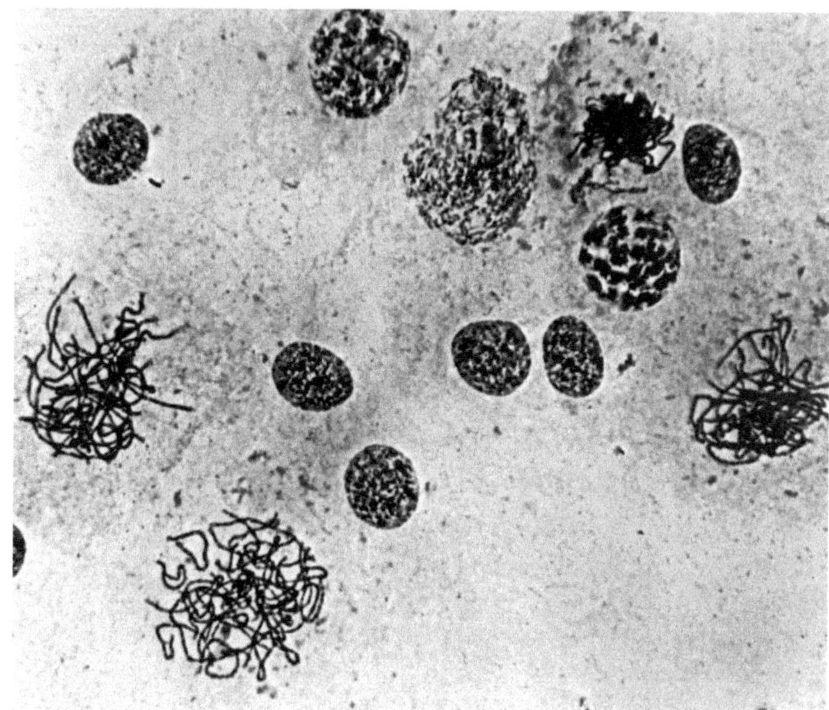

Fig. 4

## Prophase
## Preleptotene · Spiralization Phase

The so-called preleptotene initiates the prophase of the first meiotic division. It represents a phase of transition between oogonium and leptotene oocyte. In recent years it has been shown that the preleptotene is characterized by a striking change of spiralization and despiralization of the chromosomes.

Fig. 5. Preleptotene. The spiralization starts with condensation of some chromosome segments

× 2,250

Fig. 6. Preleptotene. The chromosomes continue to shorten and thicken and the condensing segments increase

× 2,800

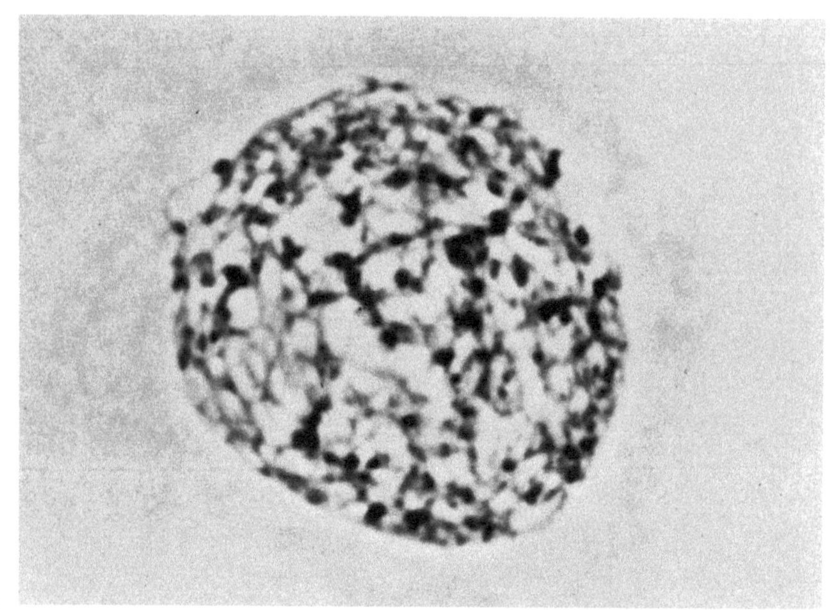

Fig. 5

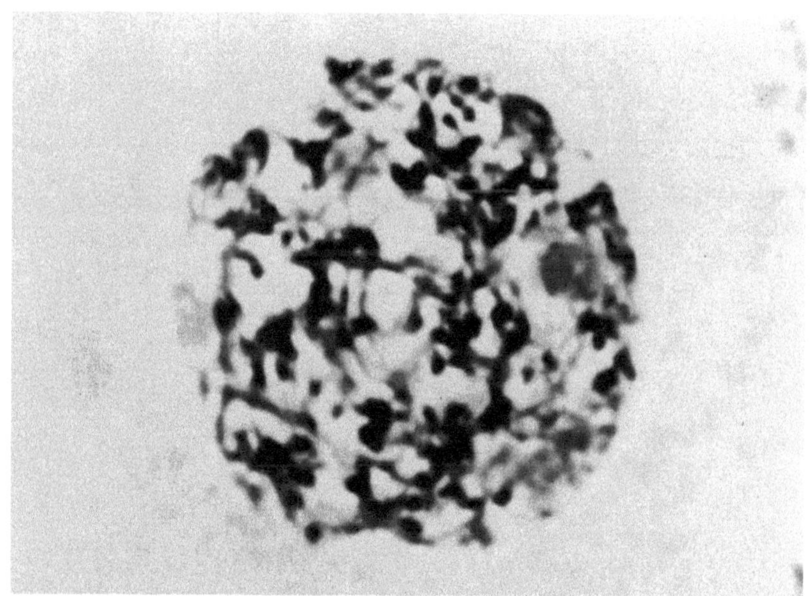

Fig. 6

## Prophase
## Preleptotene · Spiralization Phase

Fig. 7A and B. Preleptotene. Progressive spiralization and condensation of the chromosomes. In these early stages of the meiotic prophase, one to three nucleoli, sites of RNA synthesis, are already present (*arrows*)

*A*. ×2,200; *B*. ×2,280

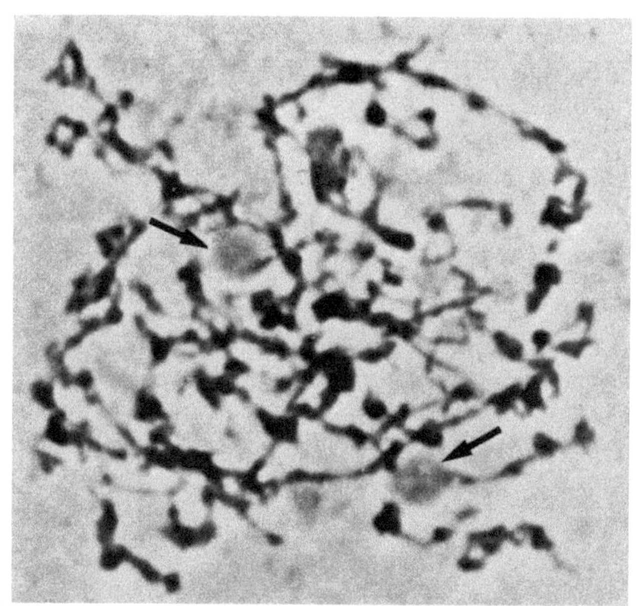

Fig. 7 A

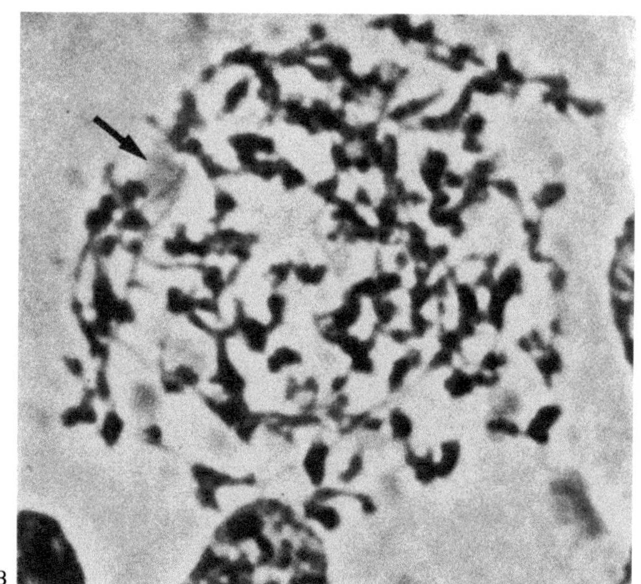

Fig. 7 B

13

Prophase
Preleptotene · Full Spiralization · Prochromosome Nuclei

Fig. 8A and B. Preleptotene. The strongly contracted and highly coiled chromosomes (*A*) condense to compact, separate chromatin blocks (*B*). Up to 46 prochromosomes are counted in nuclei with full spiralization. *Arrow* points to the nucleolus, which is in contact with several prochromosomes

*A.* ×2,700; *B.* ×2,250

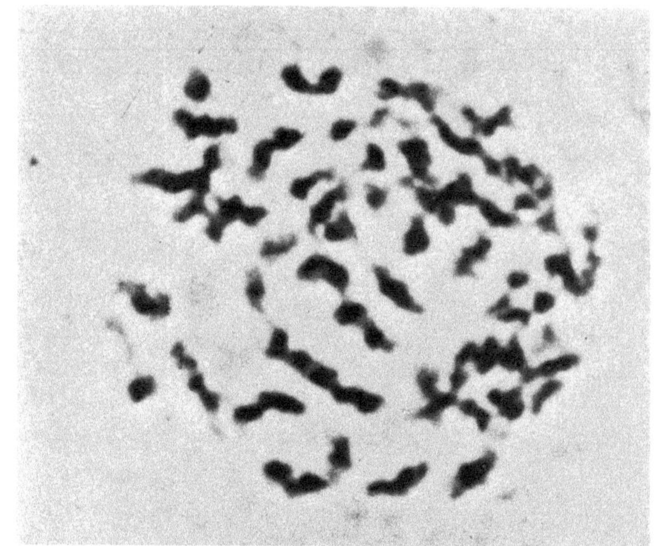

Fig. 8 A

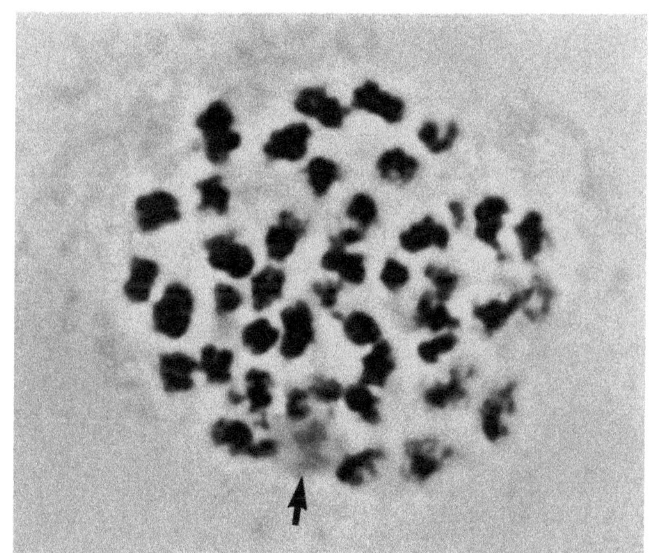

Fig. 8 B

Prophase
Preleptotene · Despiralization Phase

Fig. 9. Preleptotene. At the beginning of the despiralization phase the chromatin blocks loosen. Short processes emerge from each prochromosome

×2,250

Fig. 10. Preleptotene. The short processes are drawing out into fine filaments

×1,500

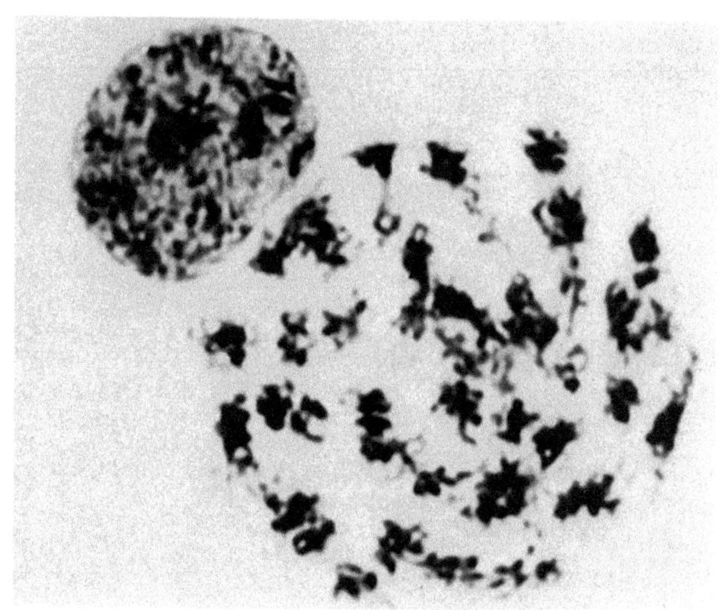

Fig. 9

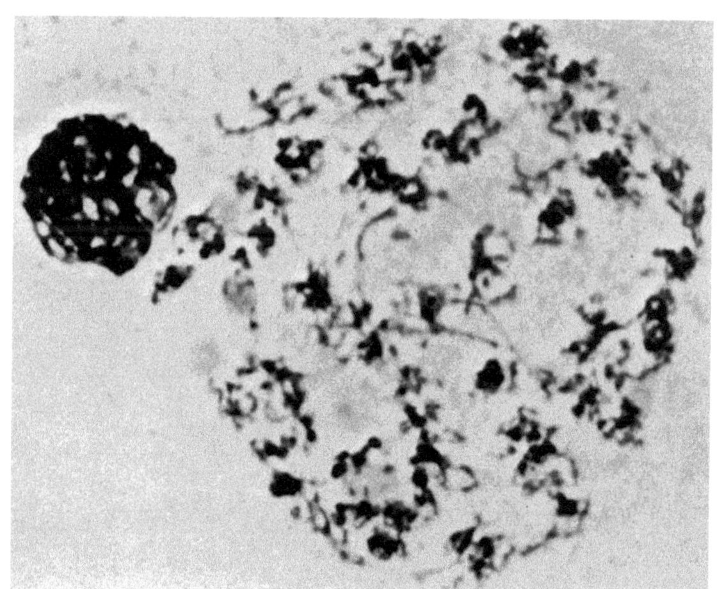

Fig. 10

Fig. 11. Late preleptotene. The chromosomes continue to despiralize

× 2,050

Fig. 12. Leptotene nucleus. The widely despiralized and extended chromosomes form a tangle of fine granulated threads

× 1,700

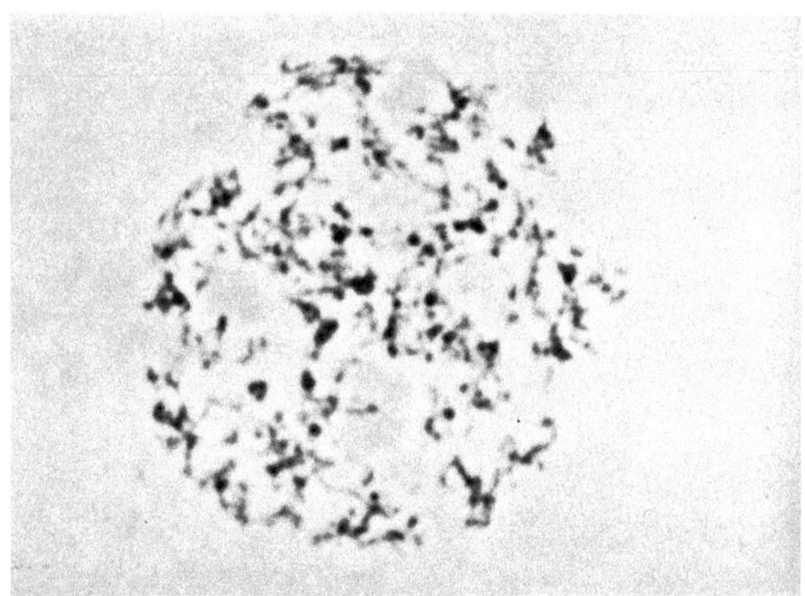

Fig. 11

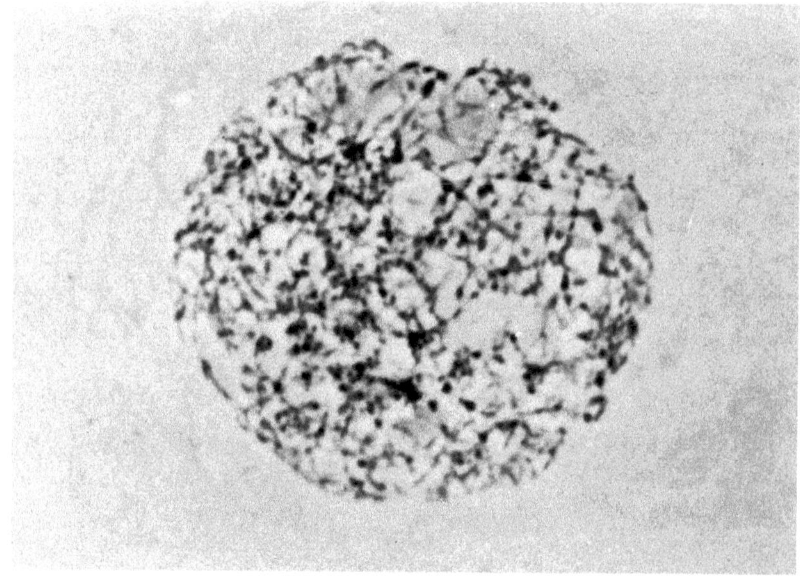

Fig. 12

Fig. 13A and B. Zygotene nuclei. In early zygotene the chromosome threads spiralize again. The homologous chromosomes — of maternal and paternal origin, respectively — begin to pair. At their ends they are often oriented toward one point of the nuclear envelope (*A, arrow*). As the polarization proceeds, the so-called bouquet stage is formed (*B*)

*A.* ×1,550; *B.* ×1,250

In late zygotene the pairing process or synapsis is completed. The homologous chromosomes are then in intimate contact along their entire length. They appear to be present in the haploid number of pairs in the following pachytene stage.

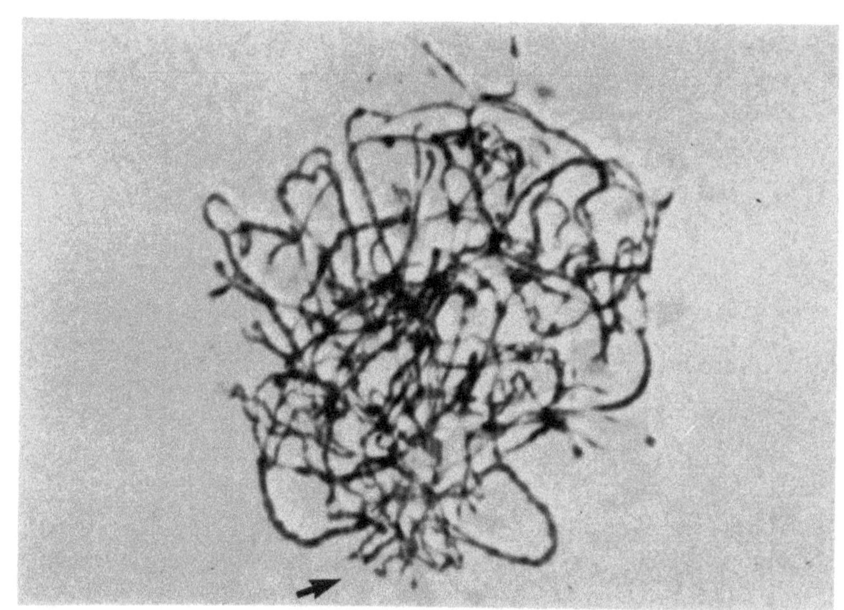

Fig. 13 A

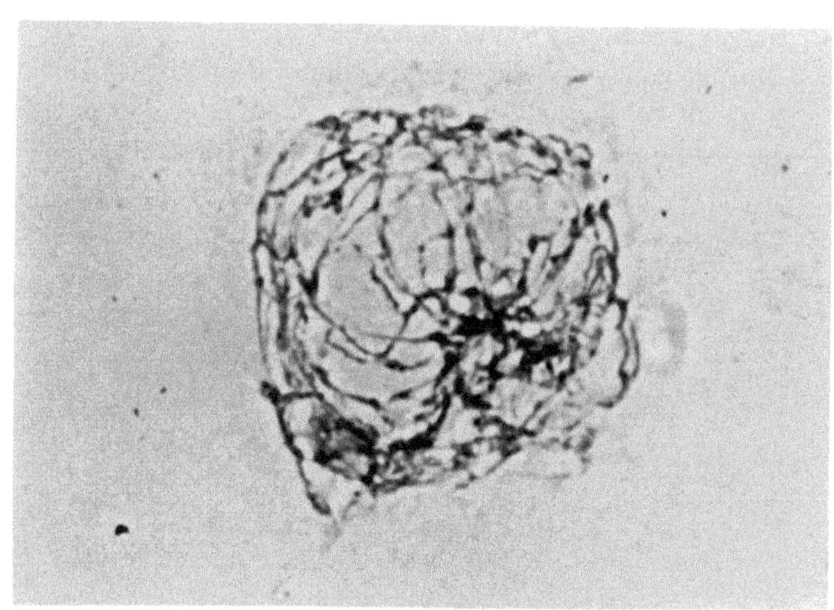

Fig. 13 B

Fig. 14. Early pachytene

× 1,800

The chromosomes, now associated in pairs by the process of synapsis, are called bivalents. Because each member of a pair consists of two chromatids, the pairing configurations are also termed tetrads. The composite structure of the pachytene chromosomes is concluded in retrospect from subsequent stages (see Figs. 16, 19, 20, and 39 A).
During pachytene the chromosomes shorten and thicken as a result of further coiling. Thus, characteristic details can be distinguished, as Figures 15 A–18 demonstrate.

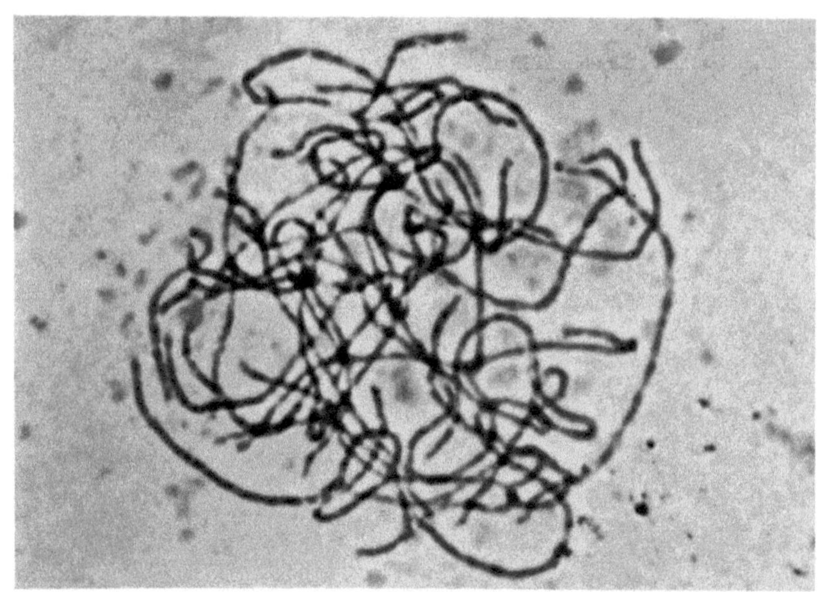

Fig. 14

Fig. 15A and B. Pachytene. *A*. Two nucleoli are present, one of them in contact with an isolated acrocentric bivalent (*arrow*). *B*. Enlargement of a portion of *A*

*A*. ×2,080; *B*. 5,400

The nucleoli are often prominent in the pachytene stage; one to two per cell are usually found. They develop in specific chromosome sites, the nucleolus organizer regions, of acrocentric chromosomes.

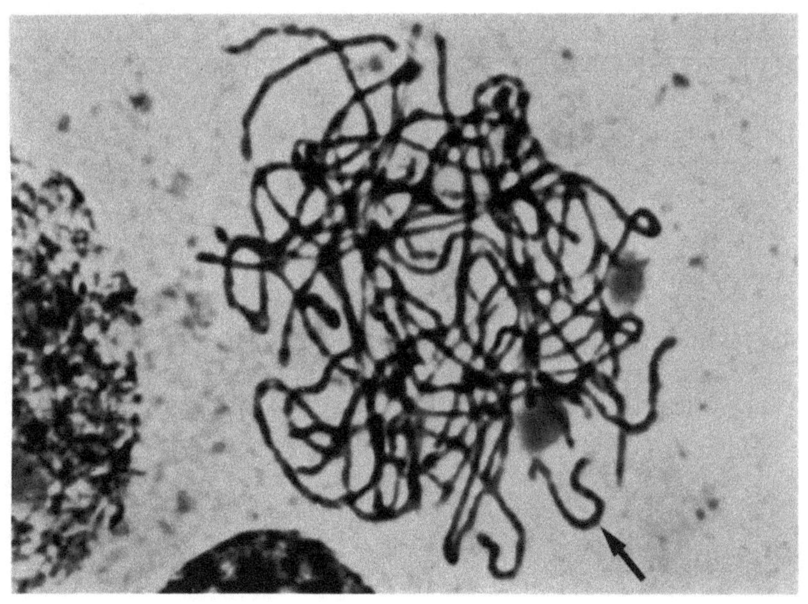

Fig. 15 A

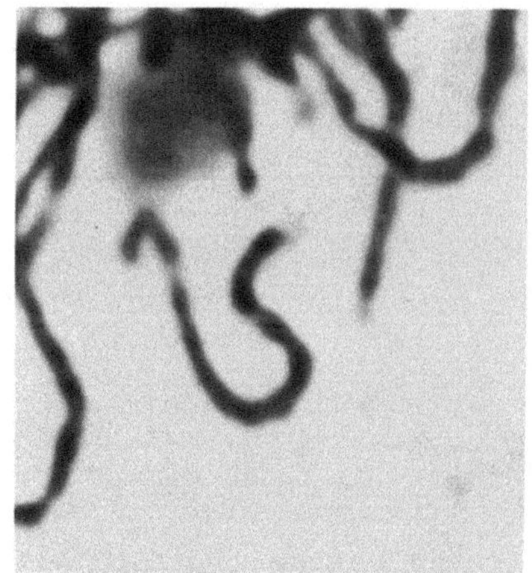

Fig. 15 B

Fig. 16. Pachytene. Several chromosome pairs are attached to one prominent nucleolus (*straight arrow*). The ends of some bivalents are untwisted, showing the two individual chromosomes (*curved arrows*).

× 2,250

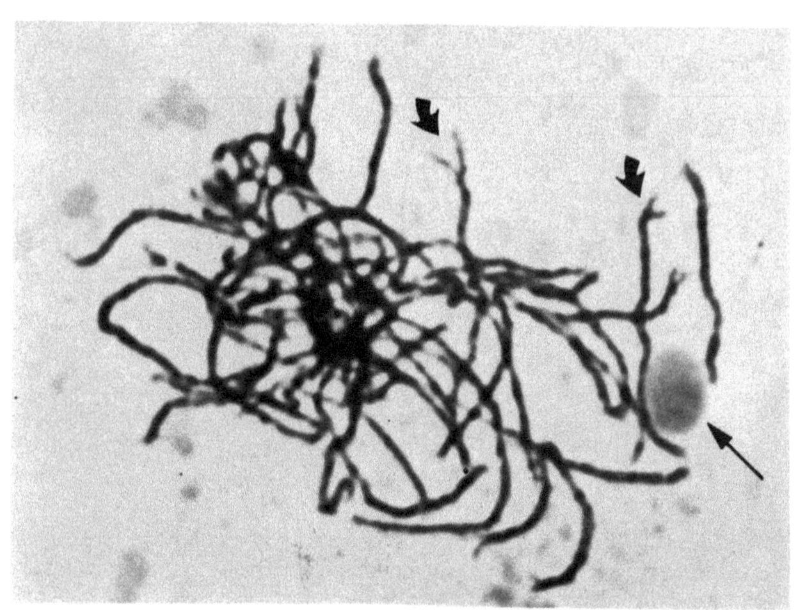

Fig. 16

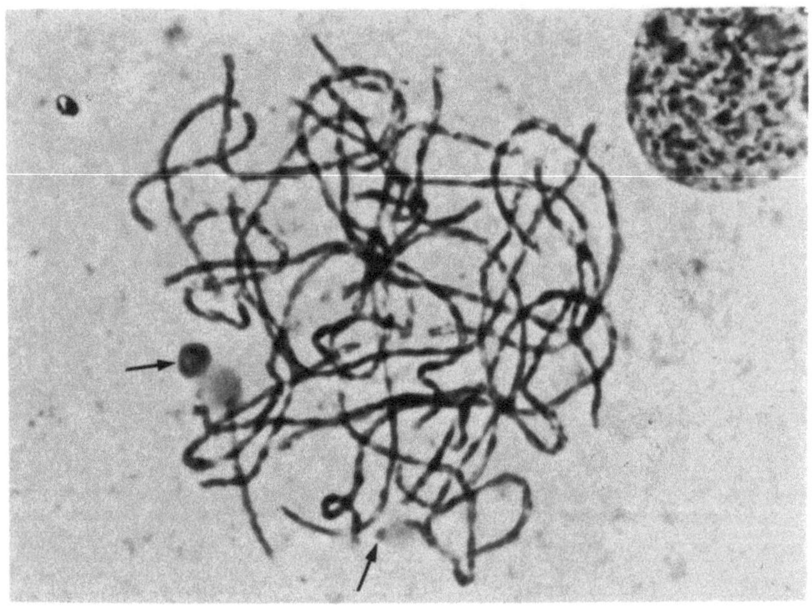

Fig. 17 A

Fig. 17A–D. Pachytene. *A.* This pachytene nucleus contains two nucleoli, conspicuous because of their "cap" (*arrows*). They were observed in prophase cells of a 22-week-old fetus, for which no more information was available. *B* and *C.* High magnification of details of *A.* The nucleoli are in contact with acrocentric bivalents, probably of the 13–15 group. *D.* Detail of another pachytene nucleus of the same fetus. Both striking nucleoli are present again and in association with several bivalents

*A.* × 1,650; *B., C.* × 5,400; *D.* × 3,240

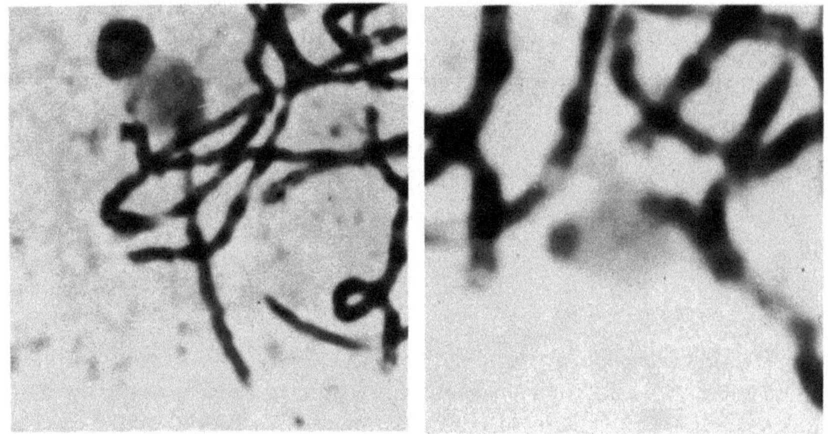

Fig. 17 B                                    Fig. 17 C

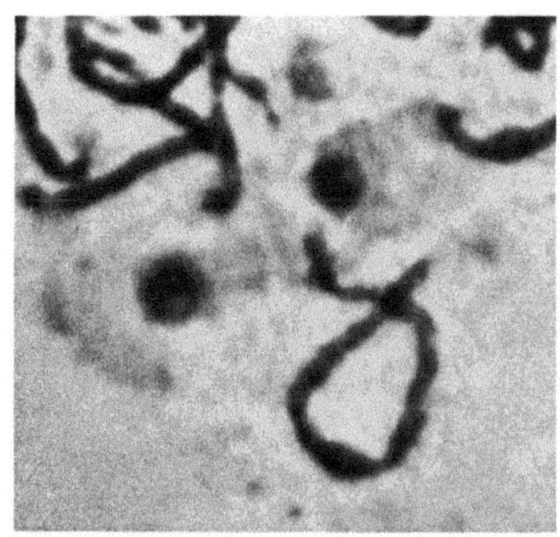

Fig. 17 D

Fig. 18. Chromomeric pattern in late pachytene. The bead-like, highly coiled and dark-stained chromomeres are arranged linearly along the chromosomes (*arrows*). Uncoiled, the chromomeres operate in the process of gene action

× 2,400

During the relatively long pachytene stage, crossing-over takes place with reciprocal exchange of genetic material between the paired chromosomes. The exchanges are observable by the formation of chiasmata in the subsequent stage, diplotene (see Figs. 19 and 20).

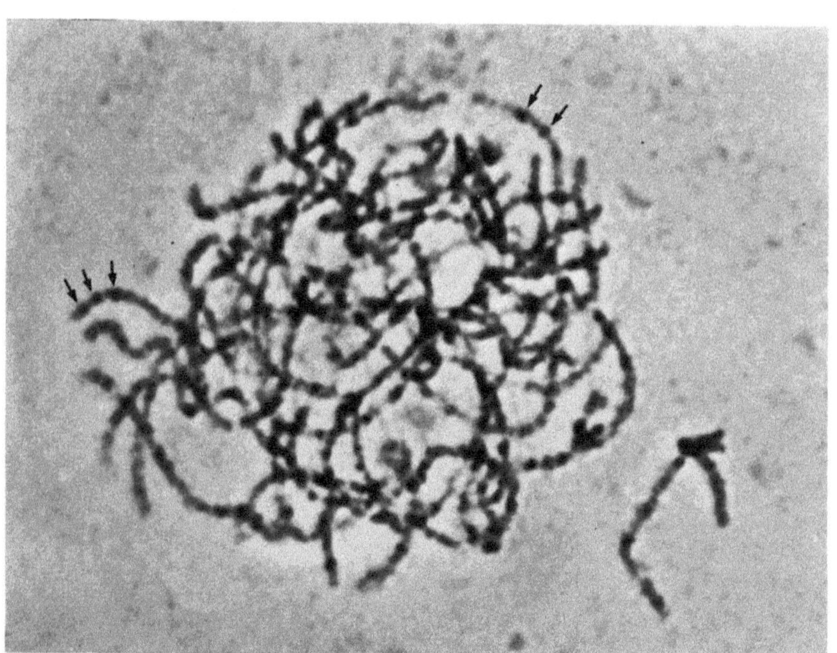

Fig. 18

Prophase
Diplotene · Despiralization · Chiasmata

In the oocyte the diplotene bivalents uncoil and become progressively longer and thinner. The pairing partners begin to repel one another. Up to their complete separation in anaphase I, they are held together by chiasmata, which represent the sites where exchange of genetic material has taken place by crossing over in the preceding pachytene stage.

Fig. 19. Diplotene nucleus. The homologues of the bivalents tend to move apart, except for their chiasmatic connections (*curved arrows*). Three prominent nucleoli are present, associated with the terminal chromomeres of acrocentric chromosomes (*straight arrows*)

× 1,650

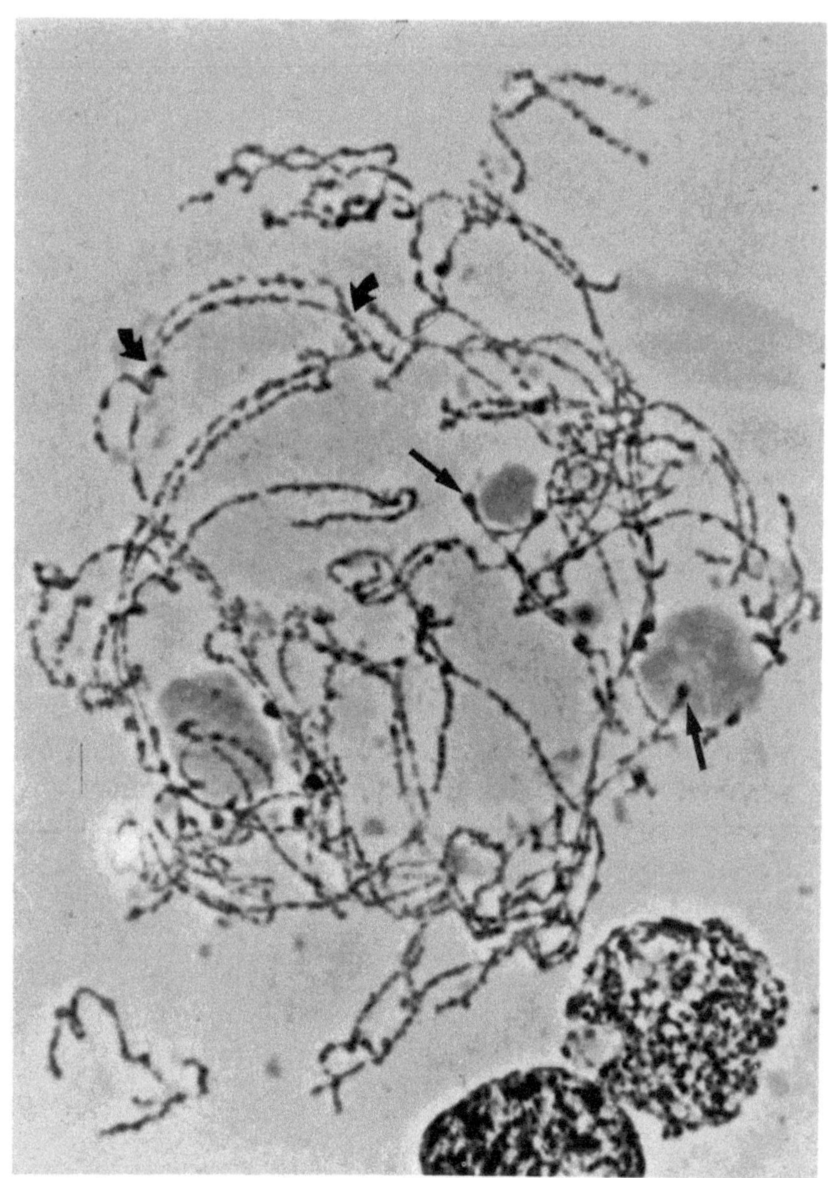

Fig. 19

Fig. 20. Diplotene. Repulsion between the homologues and chiasmata is also evident in this nucleus. Besides two voluminous and dark-stained primary, or main, nucleoli, there are several smaller-sized nucleolus-like bodies, unique in the diplotene (*arrows*). Recently it was shown that at least some of these bodies are true secondary or micronucleoli, i.e., containing rDNA

× 1,650

In late prophase the oocyte enters an arrested state of meiosis—called dictyotene—in which it will remain for many years, from birth until shortly before ovulation. During this prolonged stage, the nucleus, known as the "germinal vesicle," contains modified diplotene bivalents. Because of their characteristic appearance, as seen by electron-microscopic examination, they have been termed lampbrush chromosomes.

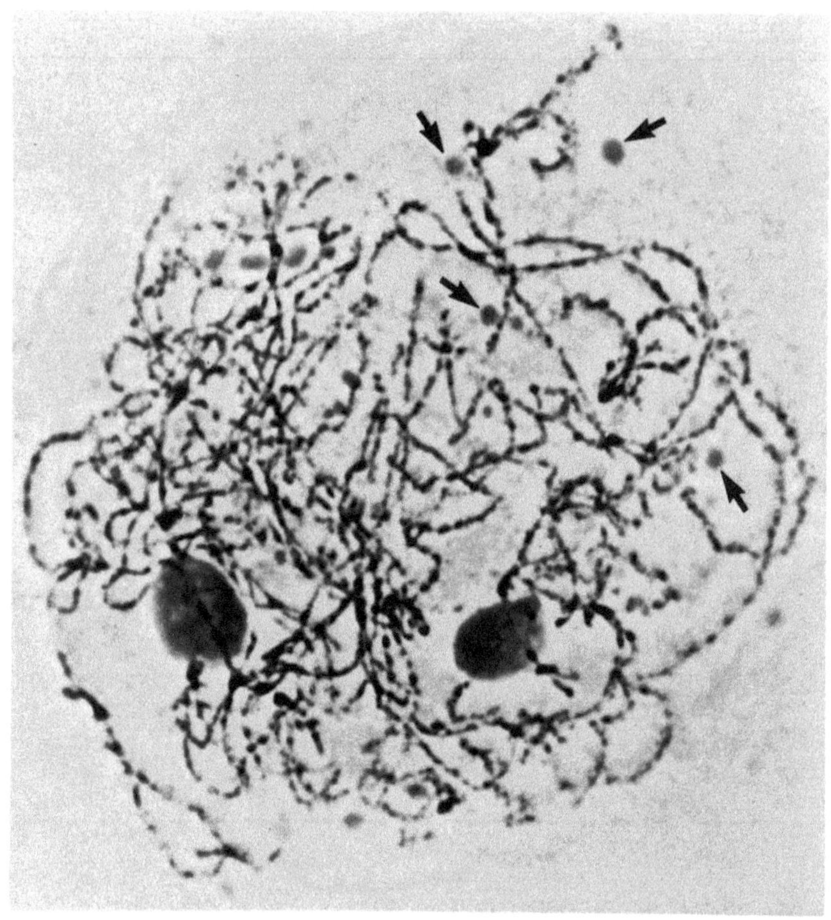

Fig. 20

# Selected References


Austin, C.R., Short, R.V. (eds.): Reproduction in Mammals. Book 1: Germ Cells and Fertilization. Cambridge: Cambridge Univ. Press 1972.

Baker, T.G.: A quantitative and cytological study of germ cells in human ovaries. Proc. R. Soc. Lond (Biol) **158**, 417–433 (1963).

Baker, T.G., Franchi, L.L.: The structure of the chromosomes in human primordial oocytes. Chromosoma **22**, 358–377 (1967).

Biggers, J.D., Schuetz, A.W. (eds.): Oogenesis. Baltimore: University Park Press; London: Butterworths 1972.

Blandau, R.J.: Observations on living oogonia and oocytes from human embryonic and fetal ovaries. Am. J. Obstet. Gynecol. **104/3**, 310–319 (1969).

Brown, D.D., Dawid, J.B.: Specific gene amplification in oocytes. Science **160**, 272–280 (1968).

Busch, H., Smetana, K.: The Nucleolus. New York-London: Academic Press 1970.

Callan, H.C.: Biochemical activities of chromosomes during the prophase of meiosis. In: Handbook of Molecular Cytology, Lima de Faria, A. (ed.). Amsterdam-London: North Holland Publishing Co., 1969, pp. 540–553.

Du Praw, E.I.: DNA and Chromosomes. New York-London: Holt, Rinehart and Winston 1970.

Eberle, P.: Die Chromosomenstruktur des Menschen in Mitosis und Meiosis. Fortschritte der Evolutionsforschung, Heberer G. (ed.) Stuttgart: G. Fischer, 1966, Vol. II.

Ferguson-Smith, M.A.: The sites of nucleolus formation in human pachytene chromosomes. Cytogenetics **3**, 124–134 (1964).

Ford, E.H.R.: Human Chromosomes. London-New York: Academic Press 1973.

Goetz, P., Suk, V., Capova, E.: First meiotic prophase in human ovaries. Folia Biol. (Praha) **22**, 25–32 (1976).

Gondos, B., Zamboni, L.: Ovarian development: The functional importance of germ cell interconnections. Fertil. Steril. **20**, 176–189 (1969).

Gondos, B., Hobel, C.I.: Germ cell degeneration and phagocytosis in the human foetal ovary. In: The Development and Maturation of the Ovary and its Functions, Peters, H. (ed.). Amsterdam: Excerpta Medica, 1973, pp. 77–83.

Henderson, S.A.: Chromosome pairing, chiasmata and crossing-over. In: Handbook of Molecular Cytology, Lima de Faria, A. (ed.). Amsterdam: North Holland Publishing Co., 1969, pp. 326–357.

Hess, O.: Chromosome structure and activity. In: Congenital Malformations, Fraser, F.C. (ed.). Amsterdam, New York: Excerpta Medica, 1970, pp. 29–41.

John, B., Lewis, K.R.: The meiotic system. Protoplasmatologia. Berlin-Heidelberg-New York: Springer, 1965, Vol. VI.

Kindred, J.M.: The chromosomes of the ovary of the human fetus. Anat. Rec. **147**, 295–311 (1963).

Langman, J.: Medical Embryology, 3rd Ed., Baltimore: Williams and Wilkins Co. 1975.

Luciani, J.M., Devictor, M., Morazzani, M.-R., Stahl, A.: Meiosis of trisomy 21 in the human pachytene oocyte. Chromosoma **57**, 155–163 (1976).

Luciani, J.M., Devictor, M., Morazzani, M.-R., Stahl, A.: Pachytene mapping of the C9 and acrocentric bivalents in the human oocyte. Hum. Genet. **36**, 197–204 (1977).

Manotaya, T., Potter, E.L.: Oocytes in prophase of meiosis from squash preparations of human fetal ovaries. Fertil. Steril. **14**, 378–392 (1963).

McDermott, A.: The frequency and distribution of chiasmata in man. Ann. Hum. Genet. **37**, 13 (1973).

Ohno, S.: Meiosis in the female. In: Perspectives in Cytogenetics – The Next Decade, Wright, S.W., Crandall, B.F., Boyer, L. (eds.). Springfield, Ill.: Charles C Thomas, 1972, pp. 37–42.

Ohno, S., Smith, J.B.: Role of fetal follicular cells in meiosis of mammalian oocytes. Cytogenetics **3**, 324–333 (1964).

Ohno, S., Klinger, H.P., Atkin, N.B.: Human Oogenesis. Cytogenetics **1**, 42–51 (1962).

Ohno, S., Trujillo, J.M., Kaplan, W.D., Kinosita, R.: Nucleolus – Organizers in the causation of chromosomal anomalies in man. Lancet **II 1961**, 123–126.

Ohno, S., Makino, S., Kaplan, W.D., Kinosita, R.: Female germ cells of man. Exp. Cell Res. **24**, 106–110 (1961).

Perry, R.P.: Nucleoli, the cellular sites of ribosome production. In: Handbook of Molecular Cytology, Lima de Faria, A. (ed.), Amsterdam: North Holland Publishing Co., 1969, pp. 620–636.

Stahl, A., Luciani, J.M.: Individualisation d'un stade préleptotène de condensation chromosomique au début de la meiose chez l'ovocyte foetal humain. C.R. Acad. Sci. Paris **272**, 2041–2044 (1971).
Stahl, A., Luciani, J.M.: Nucleoli and chromosomes: Their relationships during the meiotic prophase of the human fetal oocyte. Humangenetik **14**, 269–284 (1972).
Stahl, A., Luciani, J.M., Devictor, M., Capodano, A.M., Gagné, R.: Constitutive heterochromatin and micronucleoli in the human oocyte at the diplotene stage. Humangenetik **26**, 315–327 (1975).
Stahl, A., Luciani, J.M., Gagné, R., Devictor, M., Capodano, A.M.: Heterochromatin, micronucleoli and RNA containing body in the pachytene and diplotene stages of the human oocyte. In: Chromosomes Today V, Pearson, P.L., Lewis, K.R. (eds.) New York: Wiley, 1976, pp. 65–73.

Therman, E., Sarto, G.E.: Premeiotic and early meiotic stages in the pollen mother cells of Eremurus and in human embryonic oocytes. Hum. Genet. **35**, 137–151 (1977).

Winiwarter, von H.: Recherches sur l'ovogenèse et l'organogenèse de l'ovaire des mammifères (lapin et homme). Arch. Biol. **17**, 33–200 (1901).
Wolgemuth-Jarashow, D.J., Jagiello, G.M., Henderson, A.S.: The localization of rDNA in small nucleolus-like structures in human diplotene oocyte nuclei. Hum. Genet. **36**, 63–68 (1977).

# Oocytes from Adult Ovaries:
# First and Second Meiotic Divisions

The originally large number of germ cells, which reaches a maximum of approximately 6,000,000 during the fifth month postconception, is already being reduced in the fetal ovary because of cell degeneration; this process continues even in the adult ovary. The number of oocytes at the beginning of puberty has been estimated at 100,000, only 300–400 of which ever mature during the reproductive life span.

Under the influence of the luteinizing hormone, one oocyte at a time leaves the arrested prophase diplotene/dictyotene stage and resumes meiosis. If an oocyte is experimentally removed from its follicle and placed in a suitable culture medium, it spontaneously begins meiosis again. The progression through the various stages that lead to complete maturation is identical in vitro and in vivo: the *dictyotene stage* is followed by *diakinesis,* in which the chromosomes have already reached an advanced degree of contraction. At the end of this stage, the nuclear membrane and the nucleoli disappear and the spindle is set up. The chromosomes line up in the equator of the cell to form the *metaphase* plate *of the first maturation division.* In the subsequent *anaphase I,* the partners in the chromosome pairs separate completely from one another (disjunction) and move toward the opposite cell poles. This step is followed by *telophase I,* in which the first polar body, enclosing one-half of the chromosomes, becomes detached. The haploid secondary oocyte, with its maximally condensed chromosomes, called dyads, has now reached the *metaphase of the second maturation division.* In vivo the oocyte is ovulated in this stage and the second maturation division is completed only if fertilization occurs.

Technique and timing of oocyte maturation in vitro developed in recent years makes specific stages of meiosis obtainable for the study of chromosomes. Metaphases of the first and second divisions are essentially the stages in which examination for structural and numerical characteristics can be made.

In a metaphase I oocyte, a chromosome structural change with formation of a quadrivalent was ascertained for the first time from a patient with known balanced translocation. The findings are described in detail in this study.

From random ovary specimens, 161 oocytes in the first metaphase were observed. No numerical or structural abnormalities were detected.

An attempt was made to estimate the chiasma frequencies in early first metaphases with less condensed bivalents. The range of the chiasma count was 42–50 per cell. This agrees with the scarce data on human oocytes and is equal to or somewhat lower than that found in spermatocytes.

In some late metaphase I stages occasional G-chromosome univalents were seen, which may indicate beginning anaphase movement.

Thirty-seven randomly collected and cultured oocytes, which were allowed to progress to the second metaphase, also showed normal chromosome complements without indications of non-disjunction or errors of anaphase distribution. Two first polar bodies were available for cytogenetic analysis, giving complementary information to the second metaphases chromosomes.

The in vitro meiotic behavior of near 200 randomly collected oocytes, derived from women of a mean age of 38.3 years, provided no information on the frequency of chromosome mutations. One must keep in mind, however, that the number of oocytes usually obtained per ovary specimen represents a very small portion of the germ cell population distributed throughout the ovary.

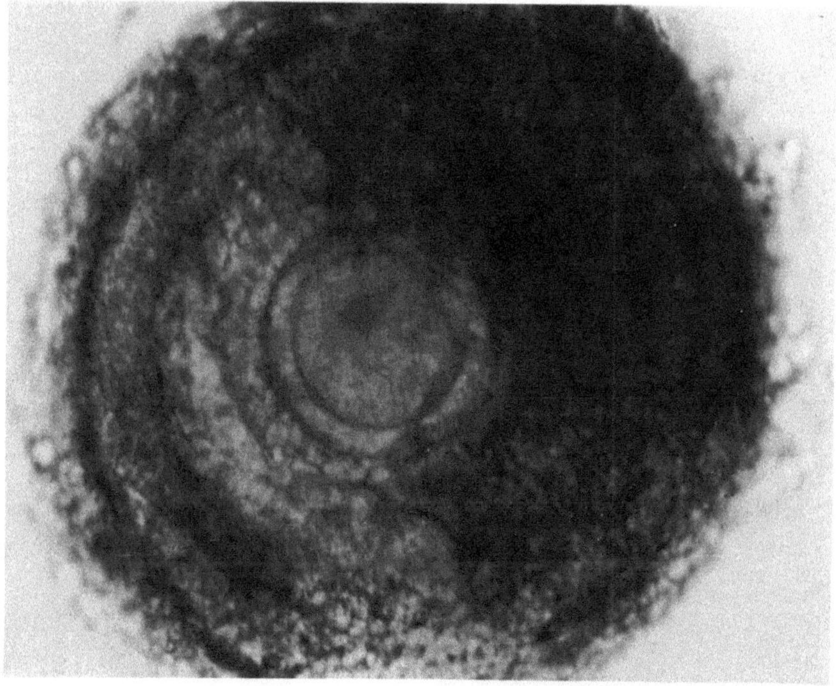

Fig. 21. Oocyte within its follicle, which was dissected intact from an adult human ovary

Unstained, ×130

Fig. 22 A and B. Living oocytes upon liberation from their follicles, normal shape. The oocytes are surrounded by dense layers of follicle cells, which form the corona radiata. In *A.*, on the *right,* the cumulus oophorus is seen, just releasing the oocyte. Both oocytes are in the dictyotene, the arrested stage of meiosis, from the time of birth (compare Fig. 30)

*A.* ×450; *B.* ×650

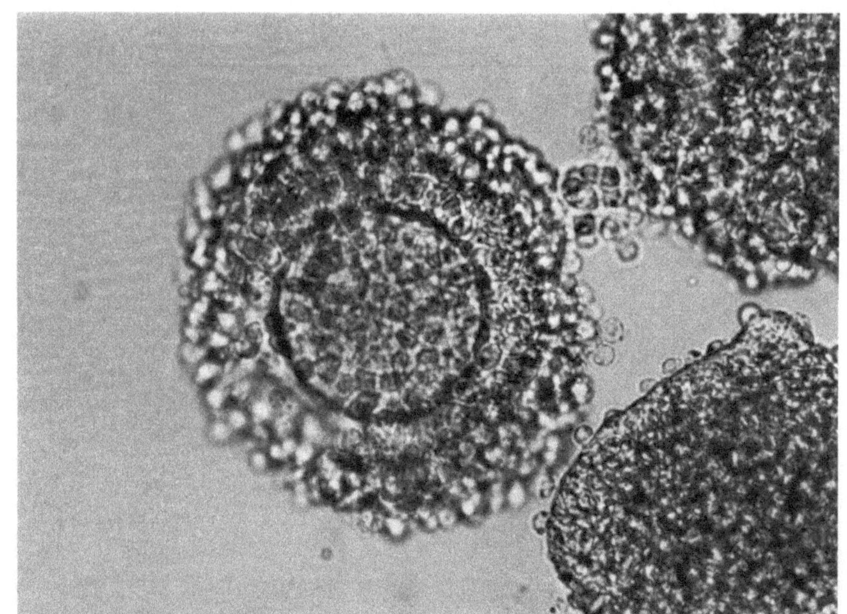

Fig. 22 A

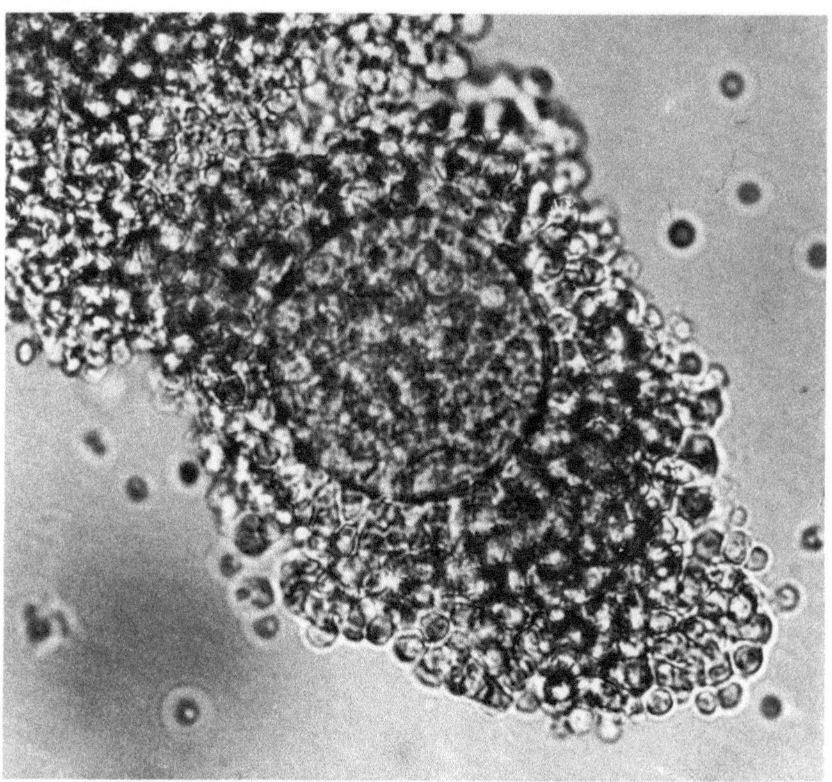

Fig. 22 B

Fig. 23. A cumulus oophorus containing two ova
×450

We have never found biovular follicles in ovaries from adult women, except in specimens of polycystic ovaries (Stein-Leventhal syndrome). Generally it is supposed that polyovular follicles are lost by atresia in late fetal or early postnatal life.

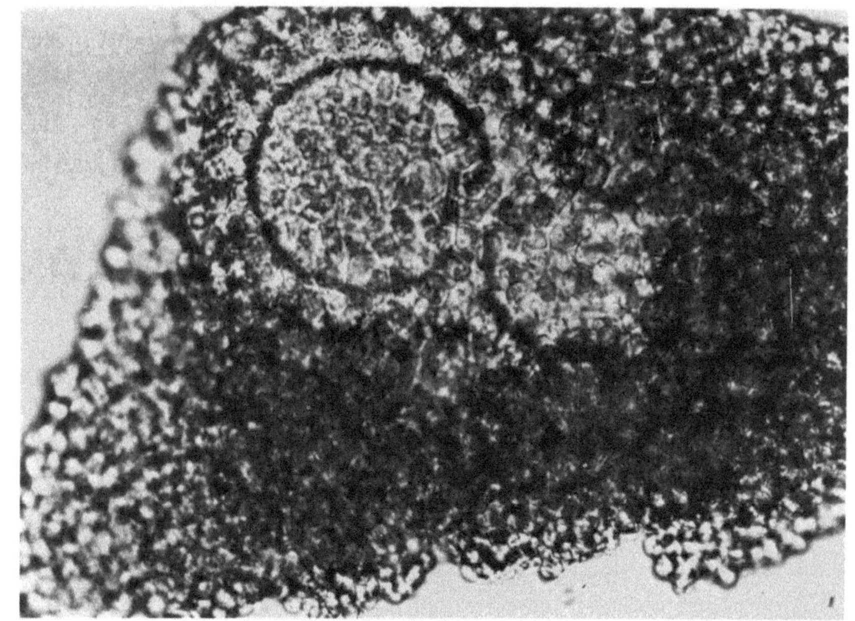

Fig. 23

47

In the human ovary the population of germ cells decreases with increasing age. The reduction in number is due mainly to the process of atresia of the follicles, during which the oocytes degenerate and are resorbed. In the age group generally available for study, the ovary specimens yield few, in the majority abnormal ova (see Figs. 24–29).

Fig. 24. Degenerate oocytes after removal from their follicles. The oocyte on the *right* exhibits loss of the corona radiata; both oocytes show shrinkage of the ooplasm

× 360

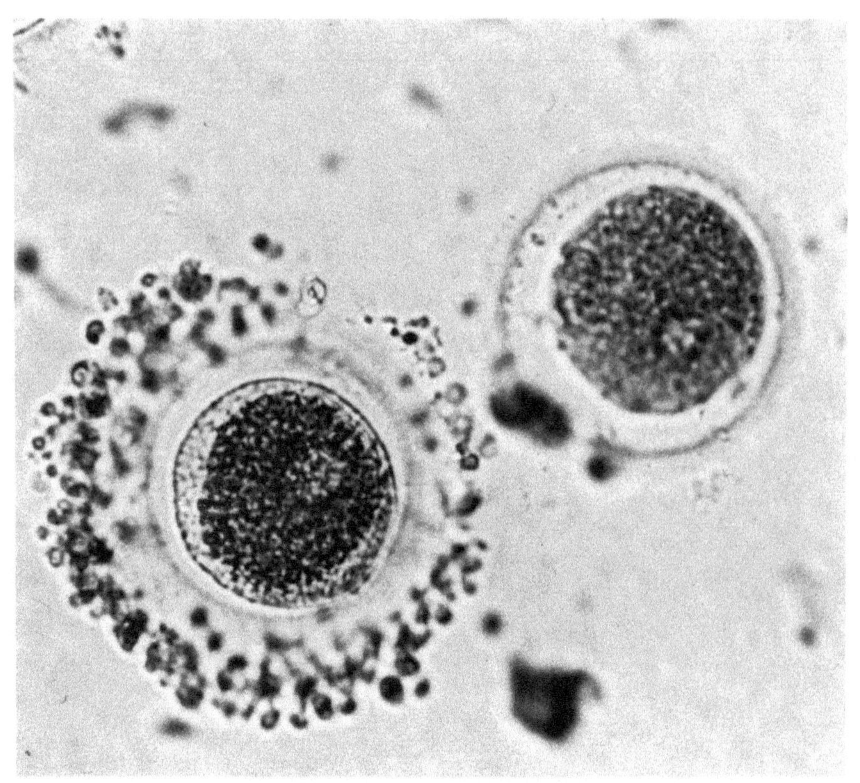

Fig. 24

Fig. 25. Oocyte, denuded of follicle cells. The ooplasm has retracted from the zona pellucida, releasing a sickle-shaped space. This type of degeneration is frequently found in atretic follicles

× 1,030

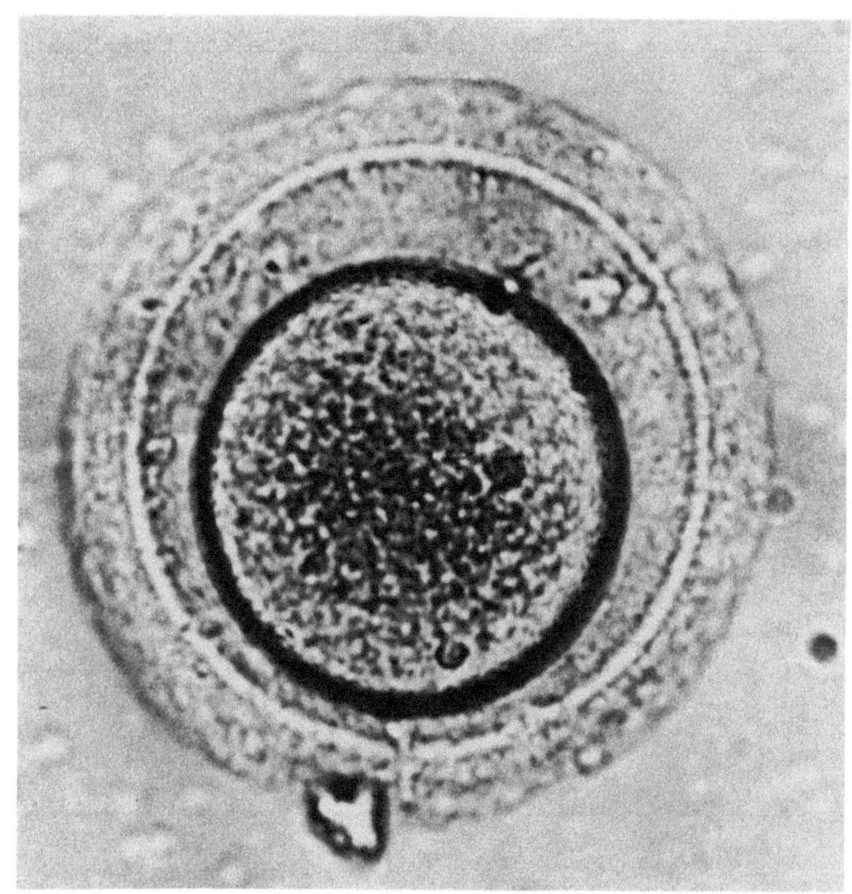

Fig. 25

Fig. 26. A grossly abnormal ovum showing severe shrinkage of ooplasm and a total loss of the corona radiata

×750

Fig. 27. Fragmentation, a form of cytoplasmic change of an abnormal oocyte. The cytoplasm has divided in two unequal portions while the oocyte was enclosed in the follicle

×870

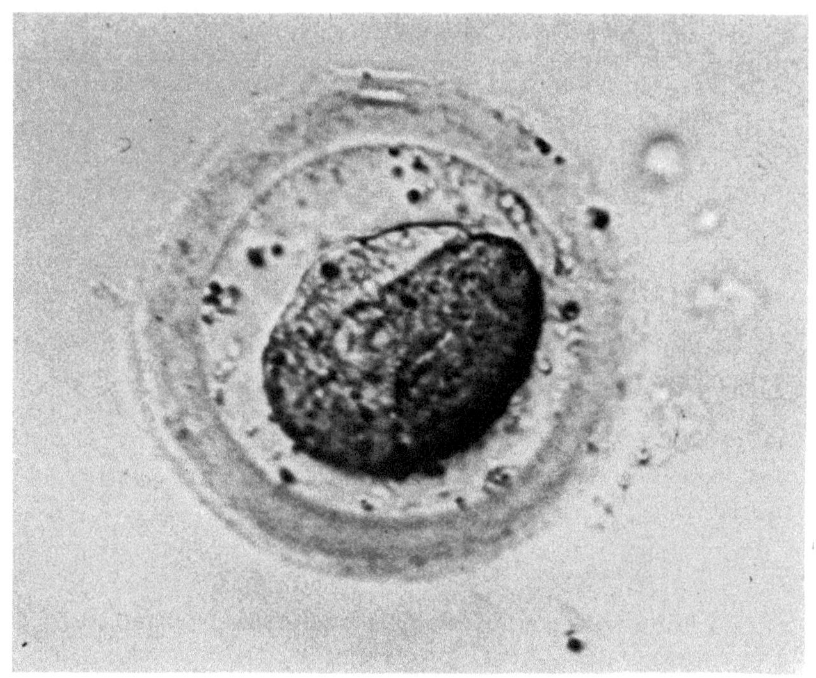

Fig. 26

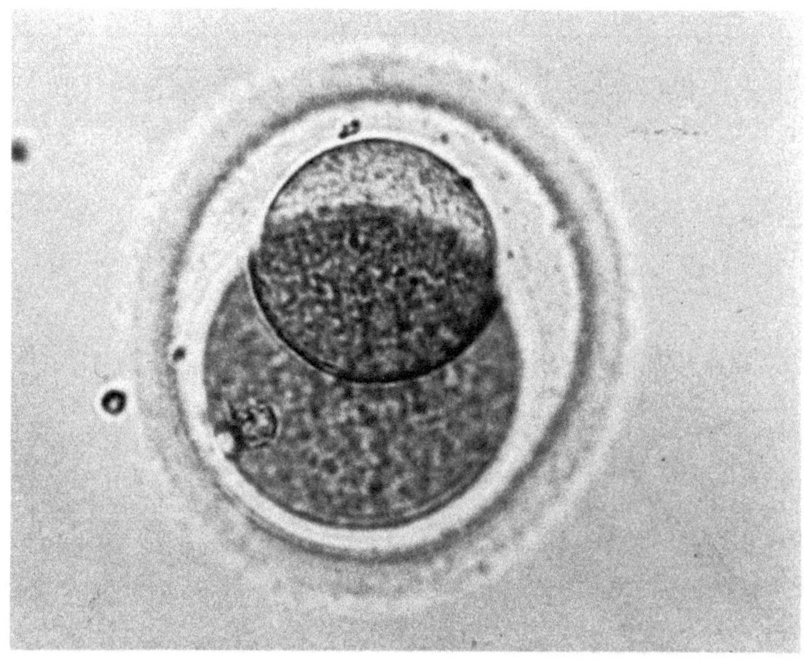

Fig. 27

Fig. 28. A fragmenting oocyte. This oocyte has undergone cytoplasmic division, apparently spontaneously, which superficially may resemble normal cleavage. In this and in Figure 29, processes are visible (*arrows*), with which follicle cells extend through the zona pellucida to reach the ooplasm. The immediately surrounding cells have a nutritive function in relation to the oocyte

× 650

Fig. 29. Another degenerative fragmentation. The cytoplasm of this oocyte is in disorganization, decaying in several unequal fragments. *Arrow* indicates a follicle cell process like that in Figure 28

× 780

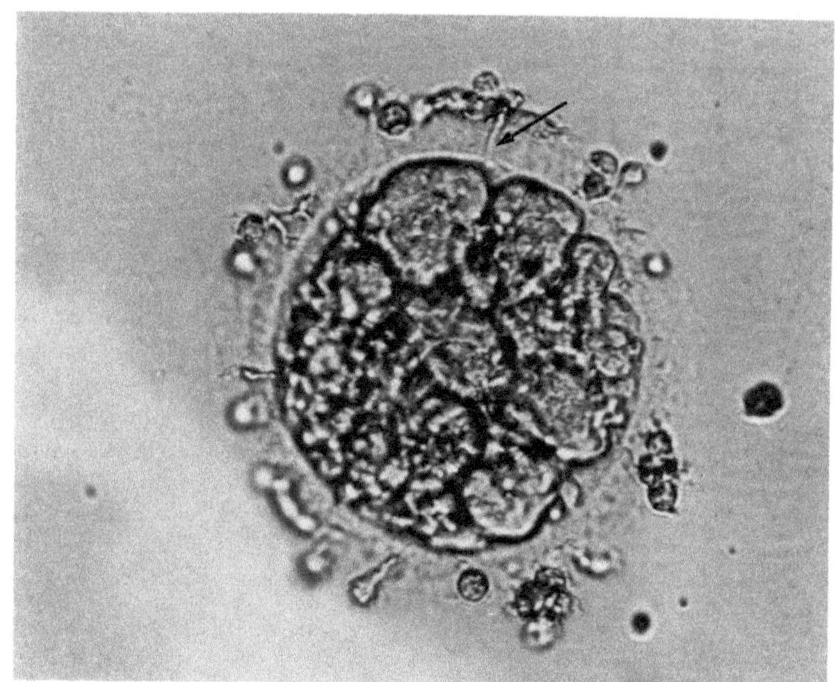

Fig. 28

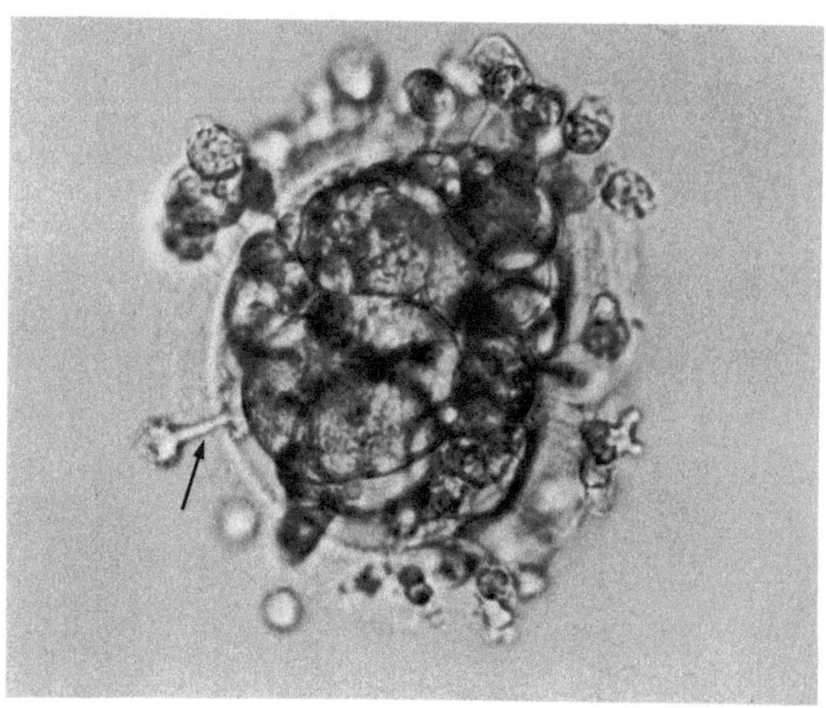

Fig. 29

Fig. 30. Living dictyotene oocyte after its release from a large follicle. Parts of the corona radiata cells were mechanically removed to show the nucleus (*g.v.* = germinal vesicle) with one highly refractile nucleolus. The oocyte is in the arrested stage of meiosis for many years. Its normal shape and size and intact germinal vesicle make it suitable for placement in a culture system, where resumption of meiosis and maturation may occur

×600

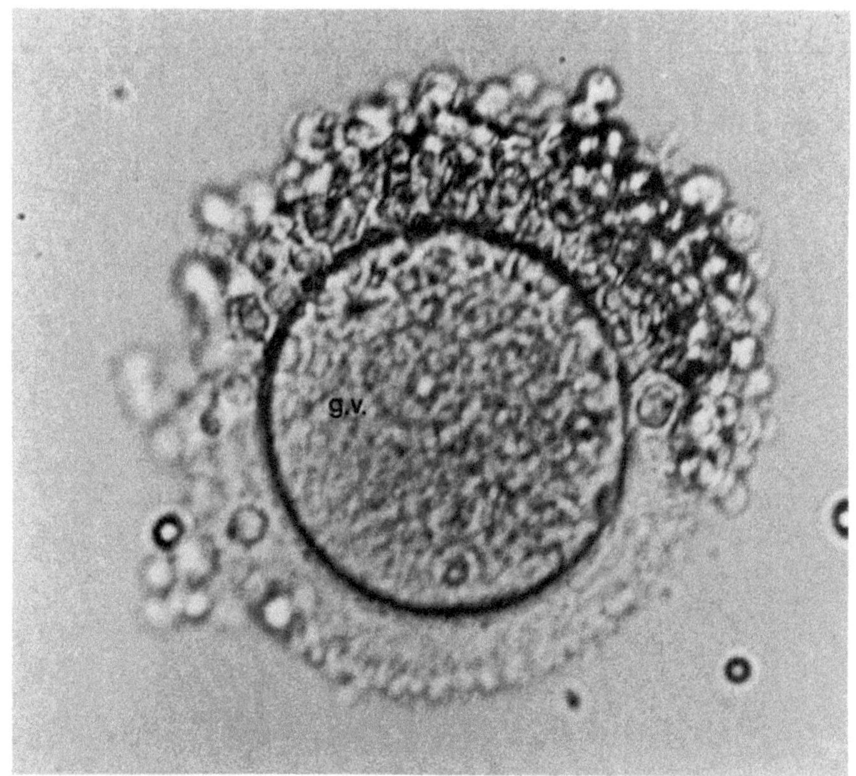


Fig. 30

57

Fig. 31 A and B. Four primary oocytes in culture. *A.* 24 h after incubation, the follicular cells of corona radiata are grown in dense layers. *B.* Enlargement of a portion of *A.*

*A.* ×120; *B.* ×400

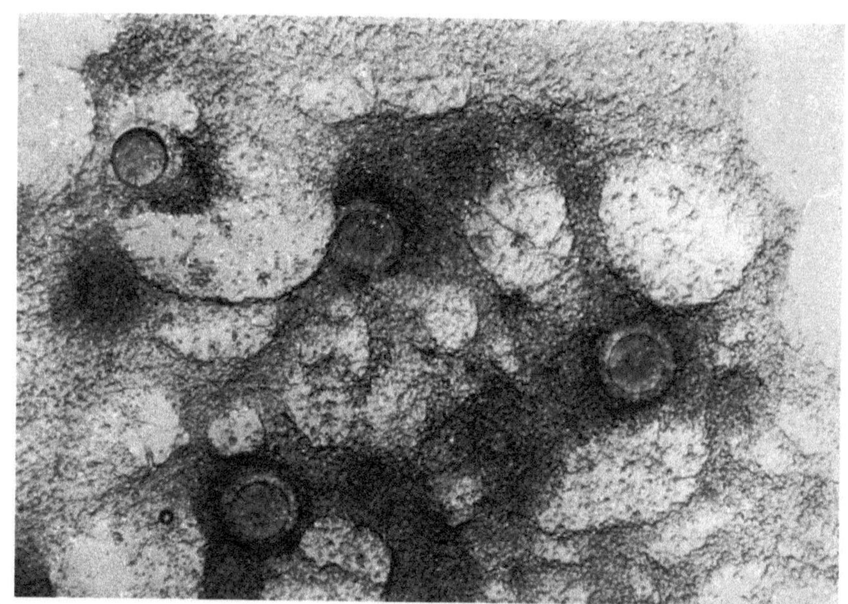

Fig. 31 A

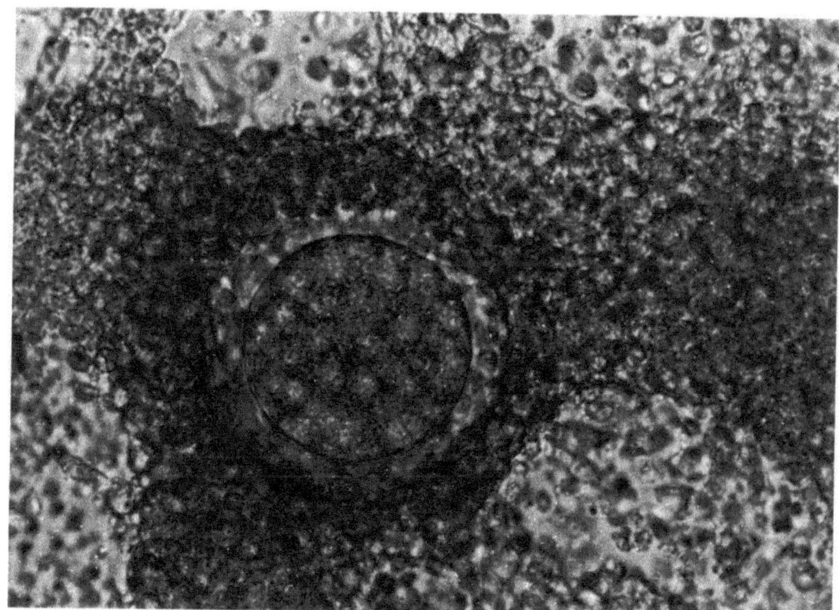

Fig. 31 B

Fig. 32. After removal of the oocyte, the follicle cells were maintained in a longer term tissue culture. Until day 7 the epithelial cells continue to grow as monolayers

× 480

Fig. 33. Subsequently, from 8 to 20 days of incubation, the confluent monolayer becomes defective, fibroblast-like cells appear, and at last the culture disintegrates with signs of necrosis

× 450

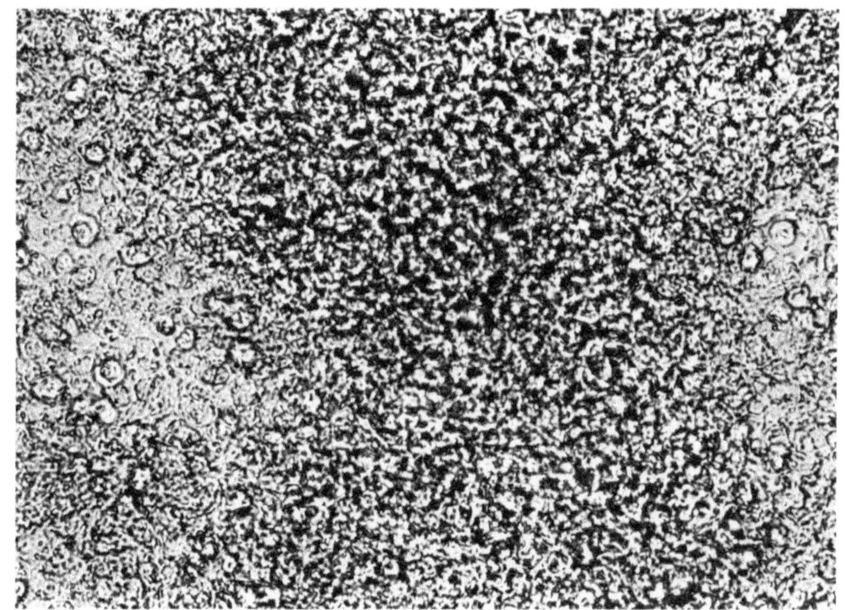

Fig. 32

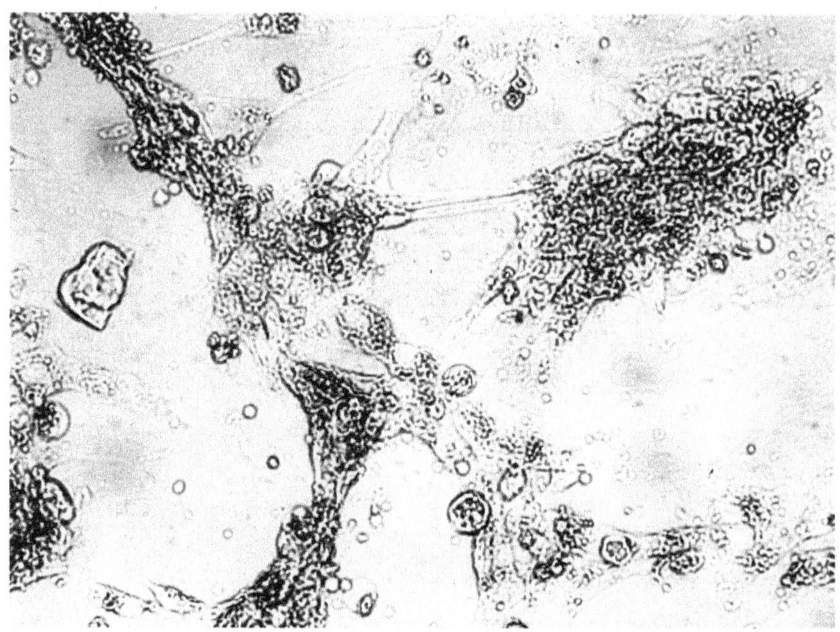

Fig. 33

Fig. 34. Anomalous fibroblast-like growth of corona radiata cells is also observed in a cultured oocyte, released from a small immature follicle. The oocyte itself seems to be degenerate

$\times 480$

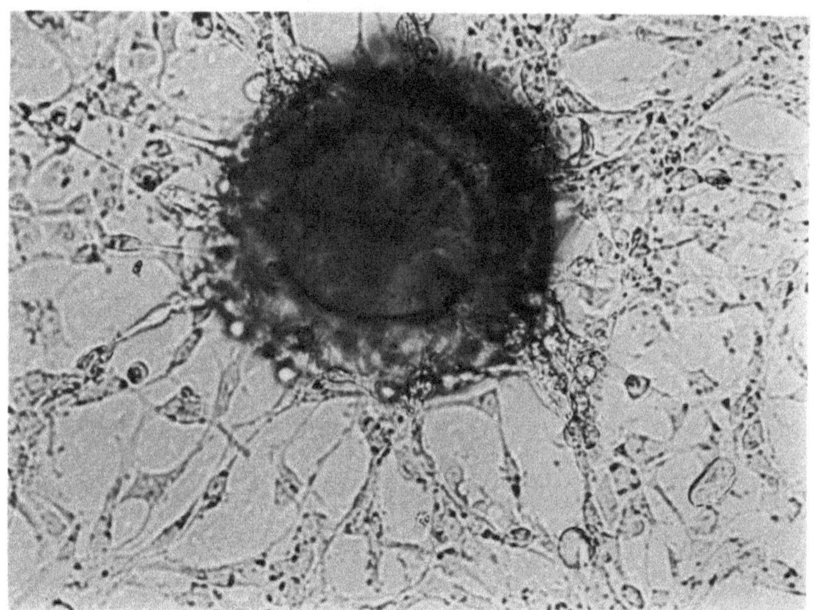

Fig. 34

In culture, the prolonged dictyotene stage is terminated, the germinal vesicle breaks down, and meiosis is resumed. Typical nuclear changes at start of culture are seen in Figures 35–38 (stained preparations).

Fig. 35. Nucleus in dictyotene stage. The circular nucleus (*g.v.* = germinal vesicle), which contains a single, dark-stained nucleolus, lies somewhat excentrically in the ooplasm

× 830

Fig. 36. A large, dense nucleolus is present, to which fine, granulated, branched filaments of chromatin are adjacent

× 1,050

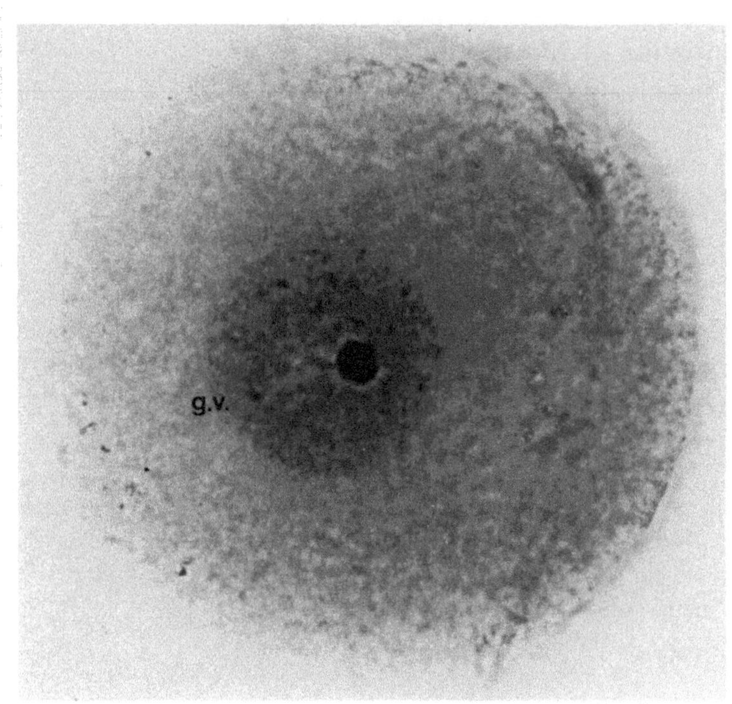

Fig. 35

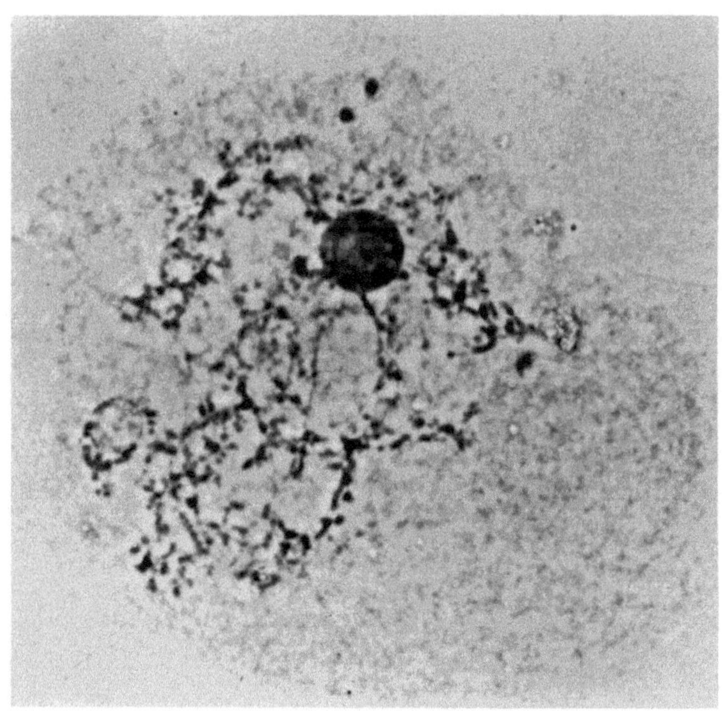

Fig. 36

## Initiation of Maturation Division
## Condensing Chromatin

Fig. 37. The chromatin has condensed around the nuclear periphery and at a higher concentration around the nucleolus

× 1,680

Fig. 38. The chromatin continues to condense, forming discrete units, i.e., bivalents, which are circularly arranged. The nucleolus, which is progressively reduced in size and soon disappears, can still be identified as a small, pale-staining body (*arrow*)

× 1,760

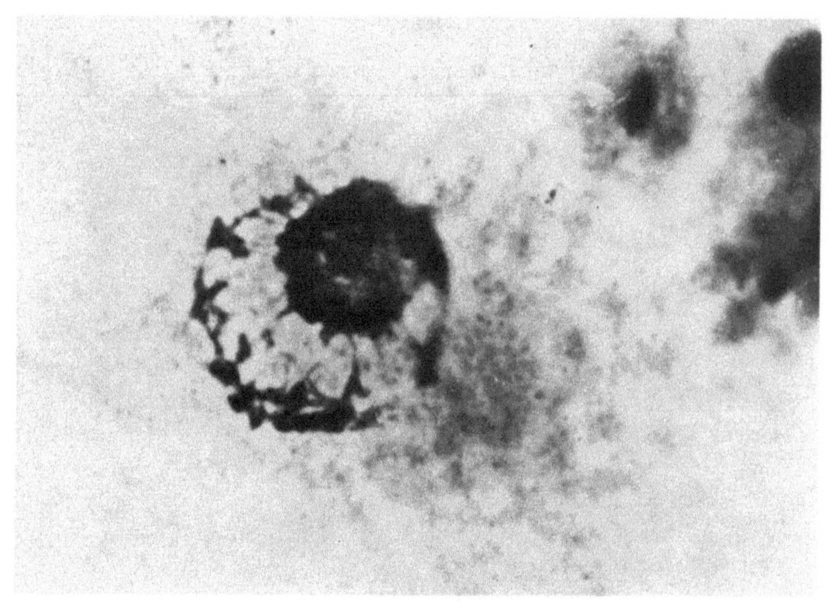

Fig. 37

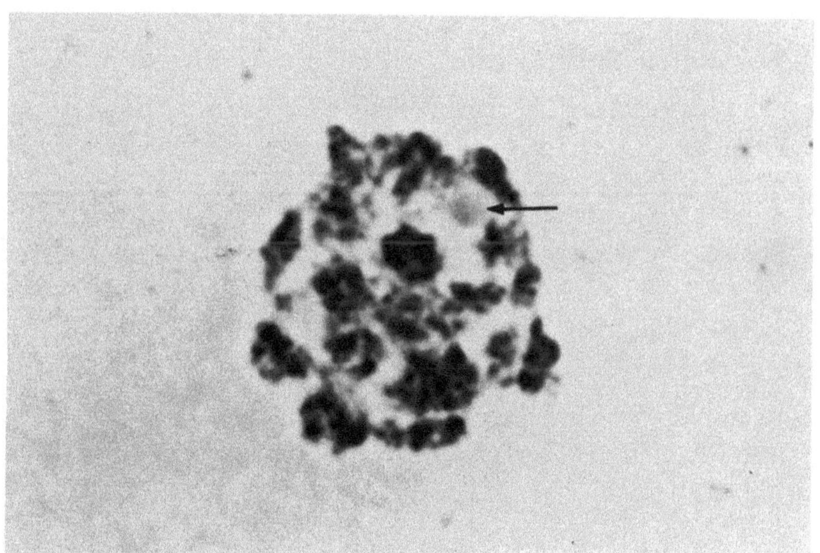

Fig. 38

Fig. 39 A and B. *A*. Bivalent or chromatid tetrad. The two chromatids of each homolog are visible and the sites of genetic exchange can be seen as chiasmata in this configuration, observed by chance. *B*. The bivalent, rarely seen so clearly in human oocytes, is included in an otherwise useless, disrupted chromosome complement of an oocyte, which may be in an early and probably brief stage of maturation, supposed early diakinesis

*A*. ×3,000; *B*. ×1,280

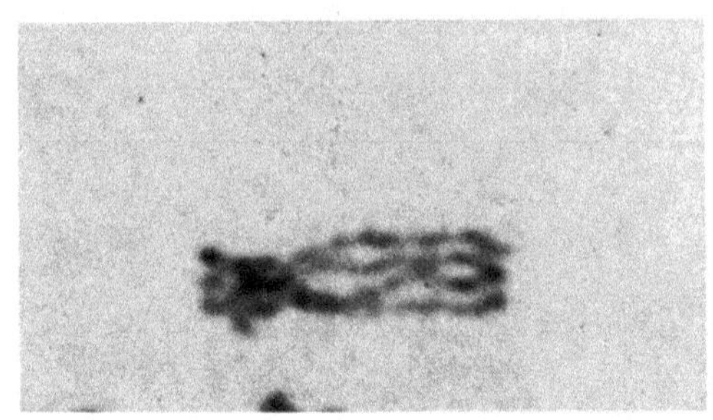

Fig. 39 A

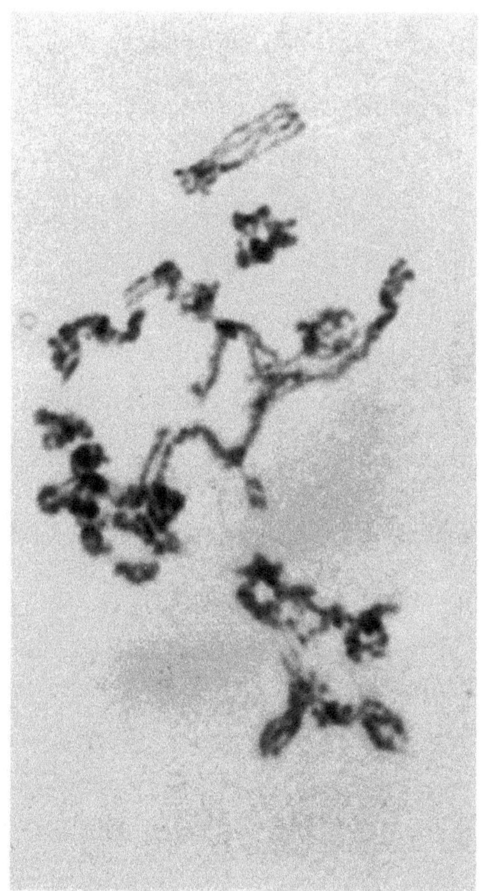

Fig. 39 B

Fig. 40. The zona pellucida of a cultured oocyte is broken up and has released the chromosomes (*arrow*). Note the proportion in size

× 330

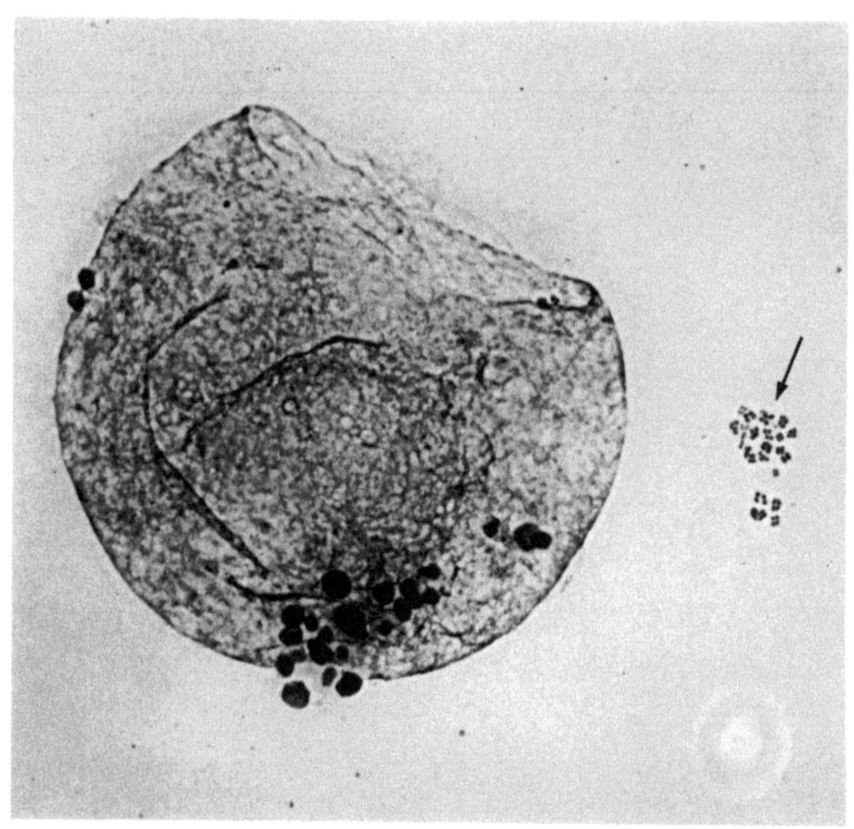

Fig. 40

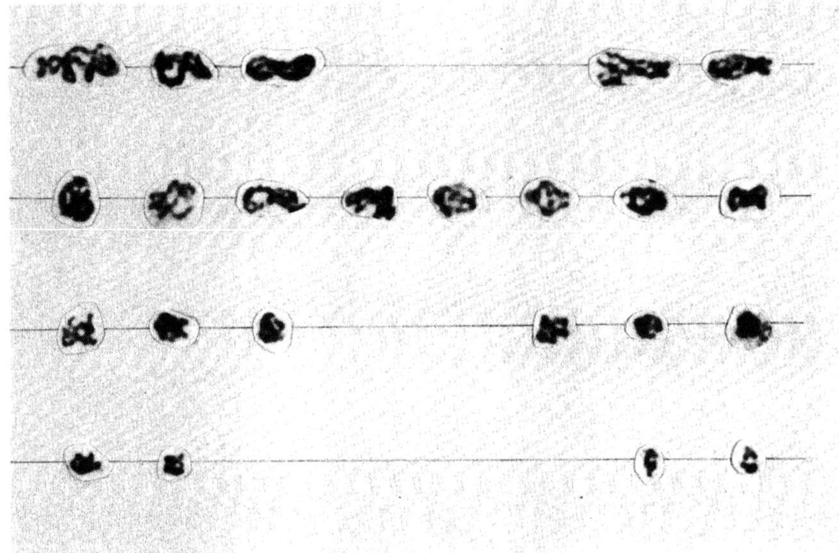

Fig. 41 B

Fig. 41 A and B. *A*. Metaphase of the first meiotic division. All the chromosomes are present as fully paired intact 23 bivalents. During preparation they were expelled at the periphery of the ooplasm. *B*. M I, 23, XX. The bivalents from *A*. arranged in a karyotype according to size

*A*. × 1,580

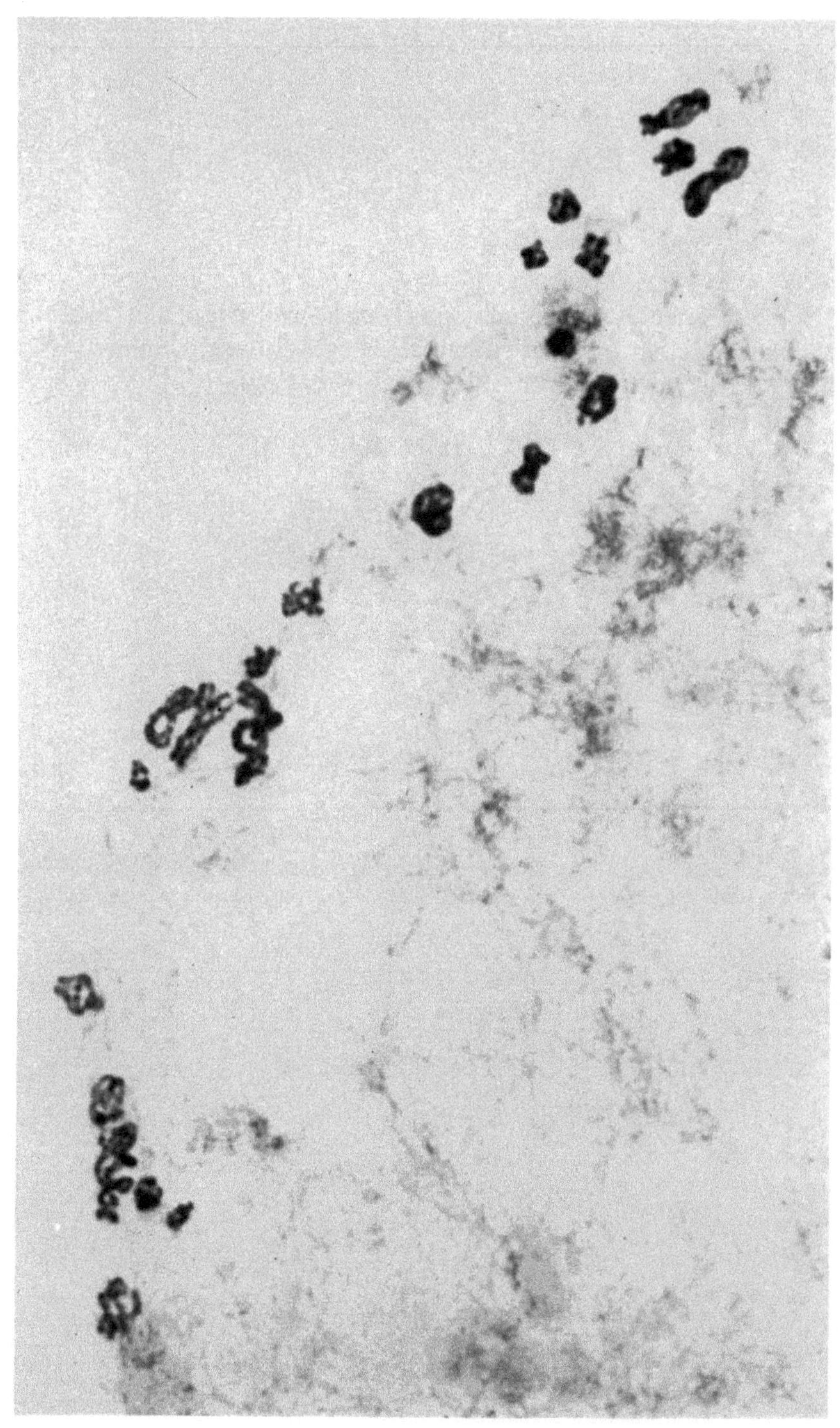

Fig. 41 A

## First Meiotic Division
## Structural Chromosome Aberration

Fig. 42 A and B. Chromosome mutation in an oocyte of a balanced translocation carrier. Complex structural changes between chromosomes 4 and 18 (*arrows*) have given rise to a quadrivalent (*Q*).

*A*. Cell in metaphase I.

*B*. Karyotype: M I, 22, XX, IV (4; 18)

*A*. ×900

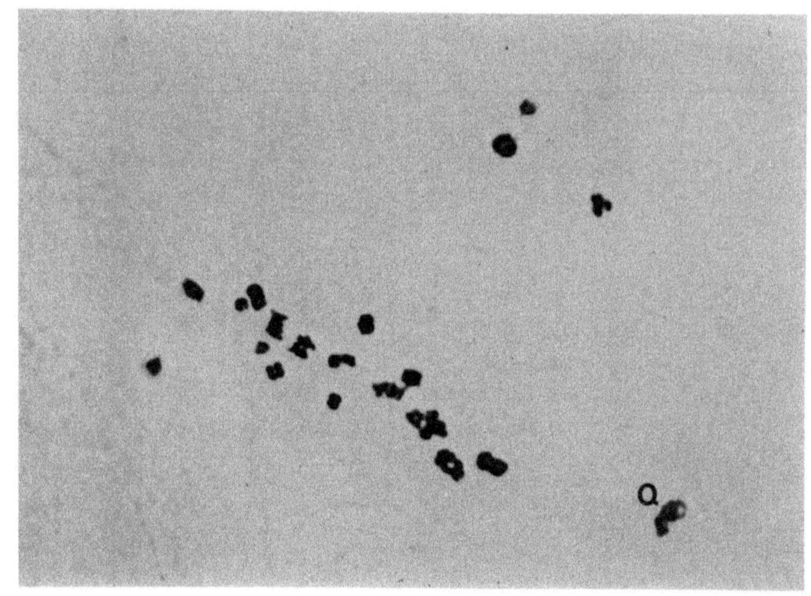

Fig. 42 A

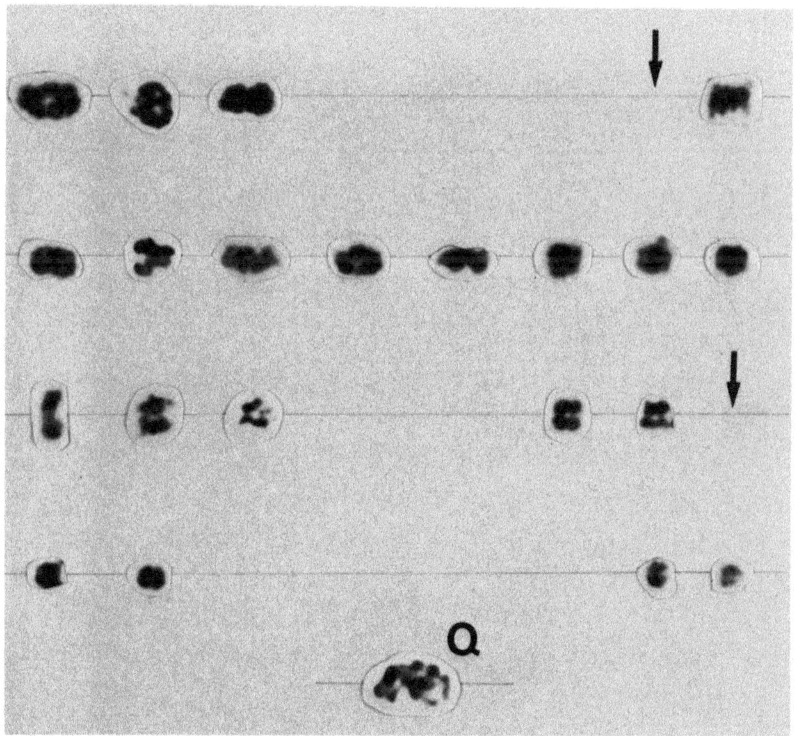

Fig. 42 B

Fig. 43. Quadrivalent, consisting of the translocation chromosomes 4 and 18 and their normal homologues. In this particular case, for which a three-break event with insertion of a fragment is assumed, synapsis involves the formation of a loop-like configuration

× 3,090

Fig. 44. Diagram of quadrivalent and formation of gametes. *Top:* The balanced carrier's chromosomes 4 and 18. *Middle:* The meiotic pairing configuration (pachytene). *Bottom:* The possible types of gametes formed after simple segregation. If crossing-over takes place, recombination gives rise to new unbalanced structures (concept of *aneusomie de recombinaison*)

Karyotypes as result of both simple segregation and recombination were observed in members of a family whose pedigree (Fig. 45) may be informative for the high genetic risk associated with complex chromosome rearrangement.

Fig. 45. Pedigree of a family with segregating complex translocation 4/18 (Courtesy of Prof. H. Knörr-Gärtner, Department of Clinical Genetics, University Ulm)

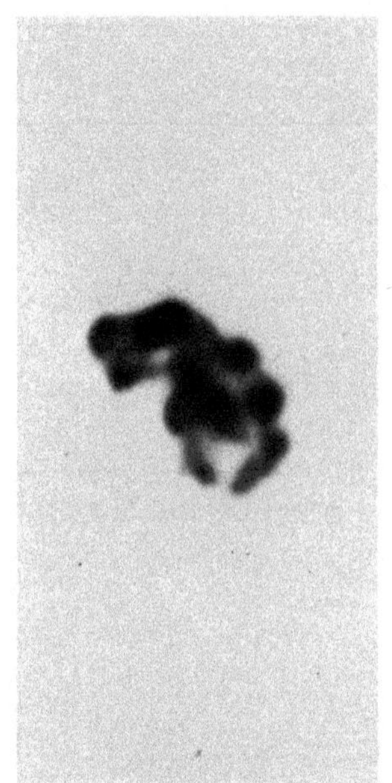

Fig. 43

Fig. 44

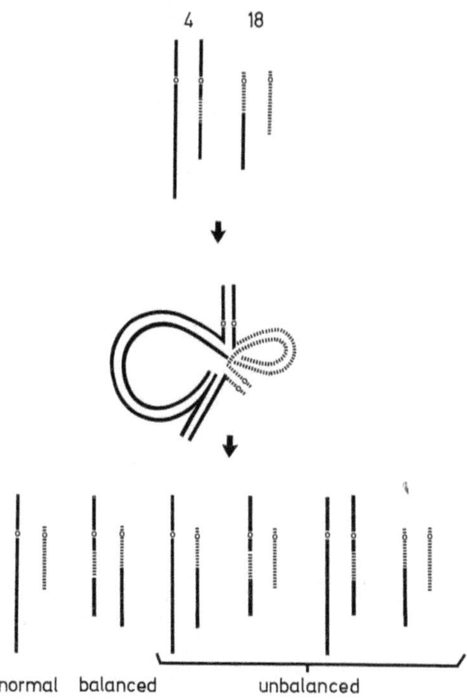


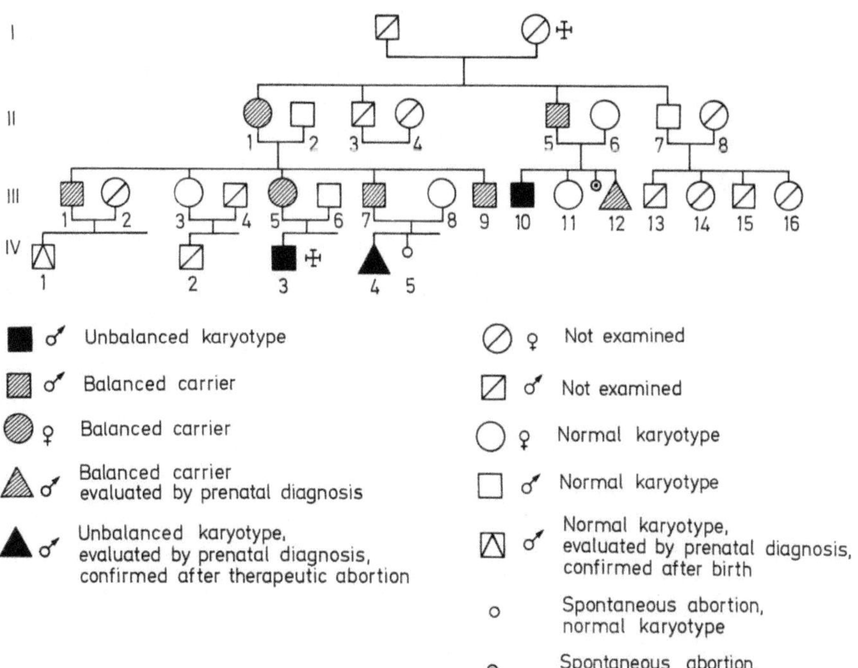


Fig. 45

Fig. 46. Late metaphase I. The bivalents have achieved a high degree of contraction. Chiasma terminalization is in progress and some of the paired chromosomes have started to separate

×2,400

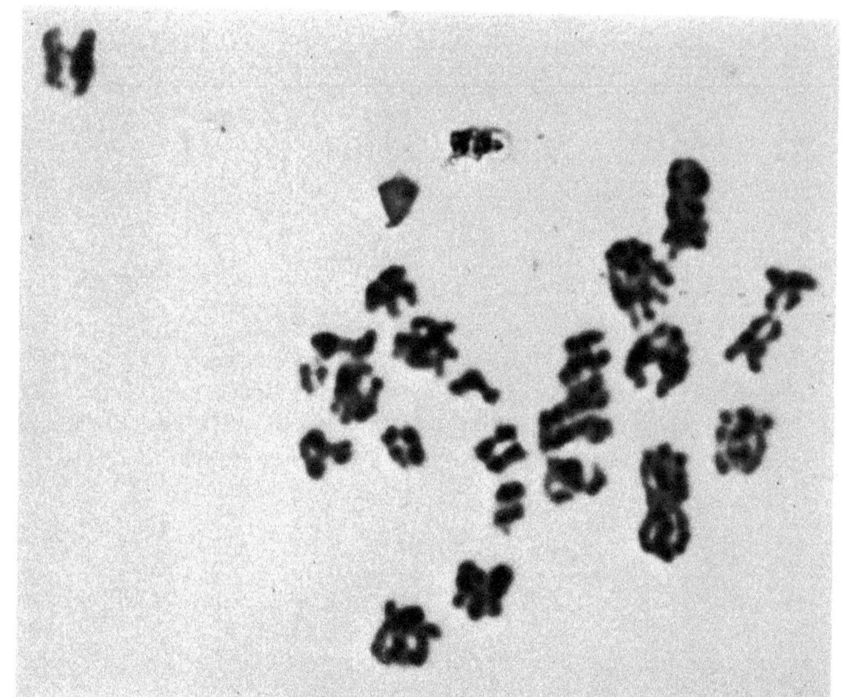

Fig. 46

## First Meiotic Division
## Types of Nuclear Degeneration

Fig. 47 A and B. Chromosomal stickiness and clumping of nuclear material. Sticky adhesions between two or more chromosomes (*A*) may interfere with chromosomal disjunction at anaphase I, whereas total clumping of chromosomes (*B*) is incompatible with viability of the cell

*A*. ×800; *B*. ×2,700

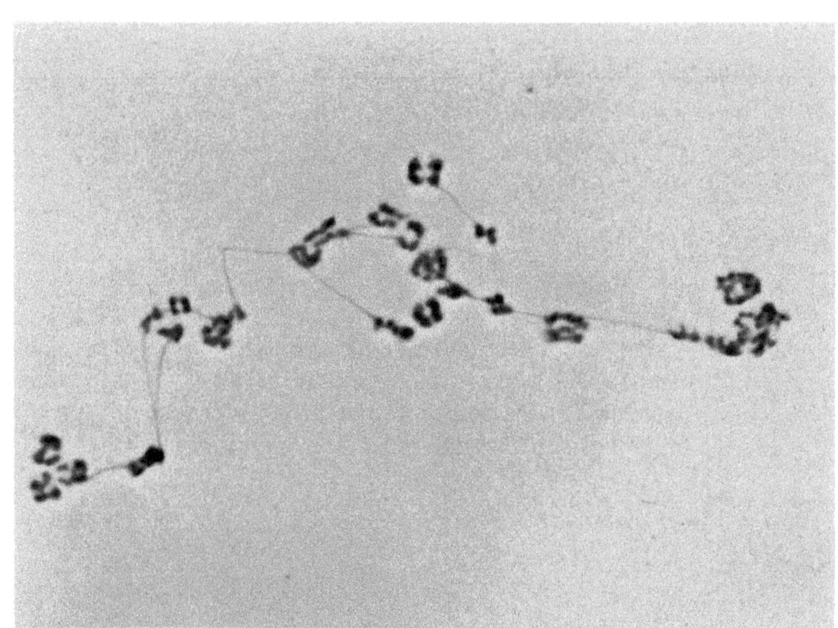

Fig. 47 A

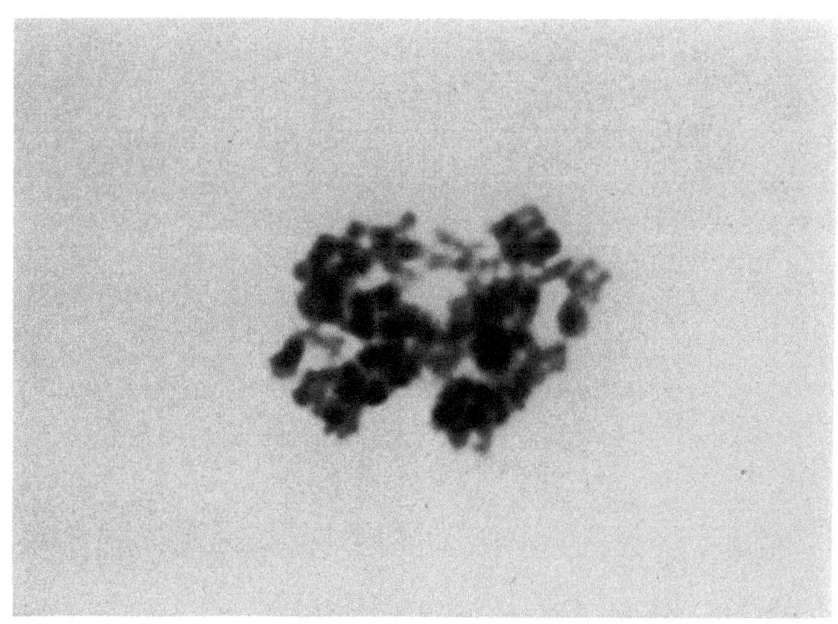

Fig. 47 B

Fig. 48. Anaphase I. The homologous chromosomes of each bivalent separate (disjunction) and move toward opposite cell poles. Thus, the reduction division is achieved: One chromosome group will remain in the oocyte and the other will pass into the polar body (see Fig. 51 A)

× 2,080

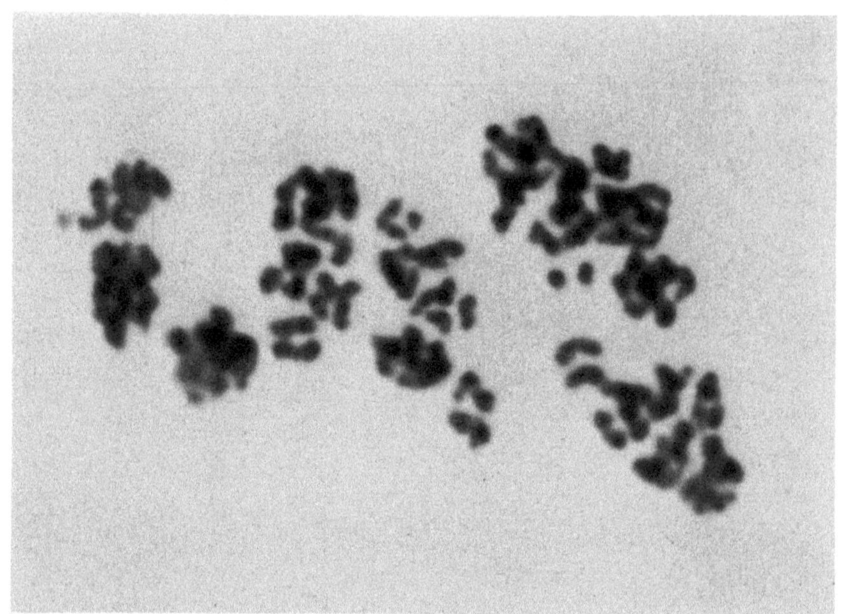

Fig. 48

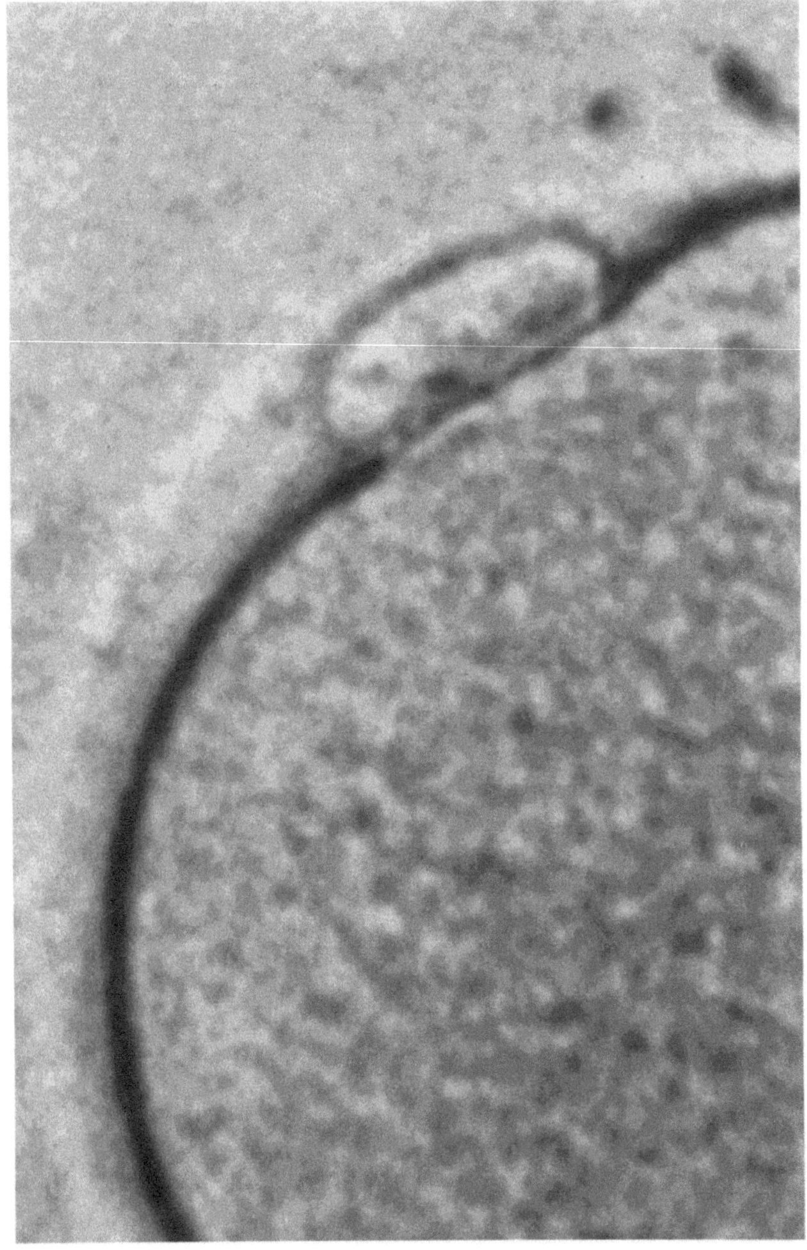

Fig. 49 B

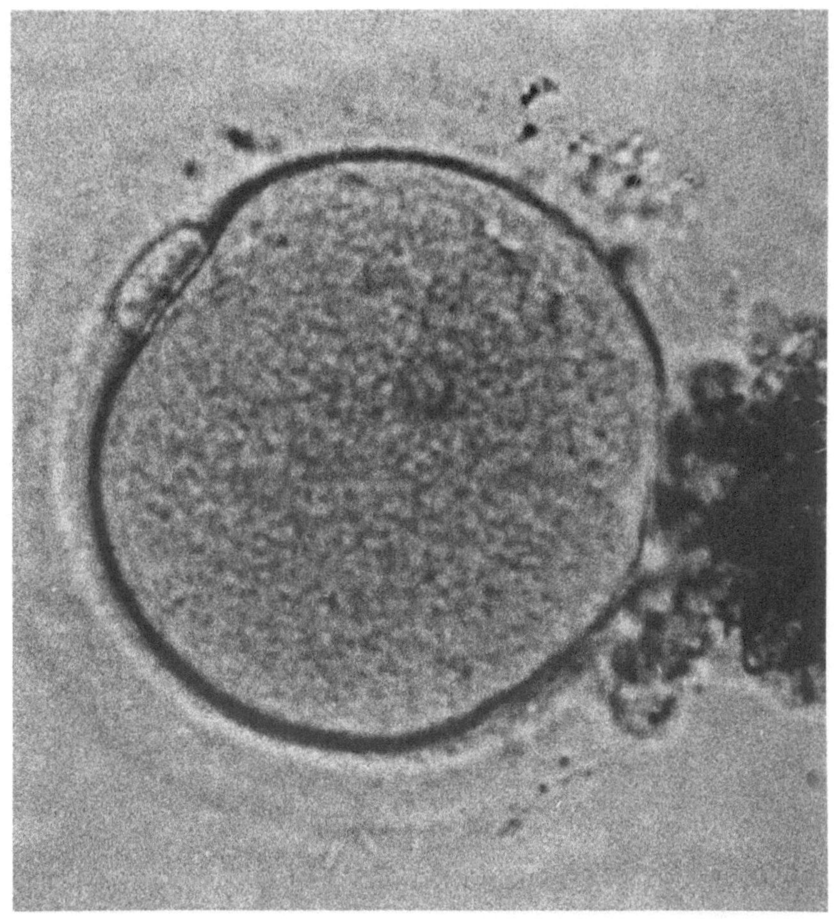

Fig. 49 A

Fig. 49 A. Living oocyte, matured in vitro. The cytoplasm has divided unequally, giving rise to the large secondary oocyte and the small first polar body

× 1,130

Fig. 49 B. The first polar body, detaching from the oocyte, lies in the small space between the vitellus and zona pellucida. Enlargement of Fig. 49 *A*

× 2,800

Fig. 50. Oocyte after culture. In order to prepare the chromosomes, oocyte and first polar body—the latter visible as a bright circle at the upper left—were caused to swell by hypotonic treatment (see Fig. 51A)

× 880

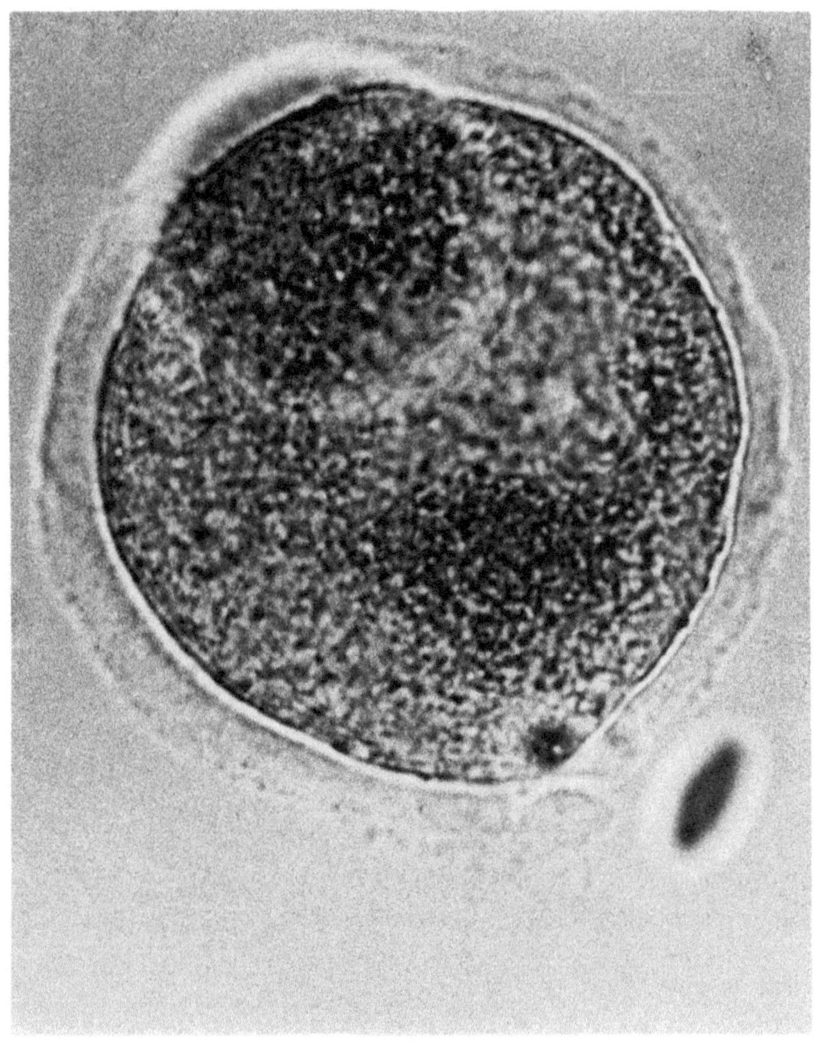

Fig. 50

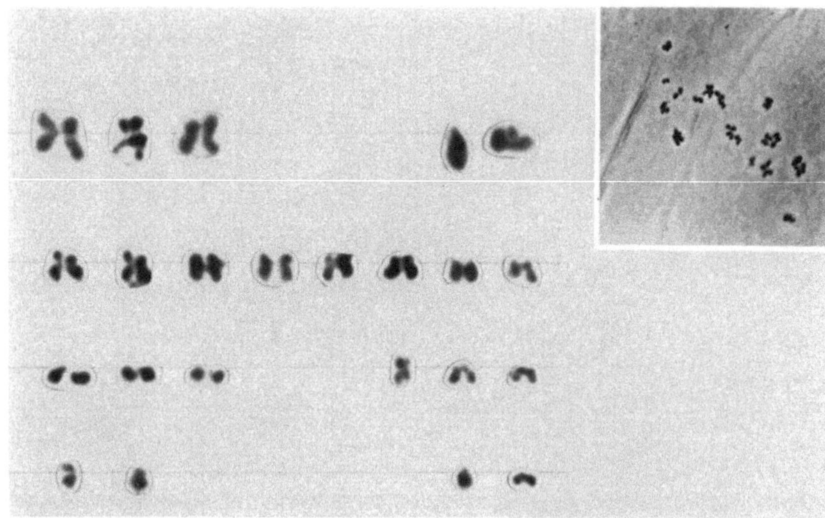

Fig. 51 B

Fig. 51 A and B. *A*. Metaphase of the second meiotic division. There are two chromosome groups: the second metaphase chromosomes of the secondary oocyte (*below*) and the scattered first polar body chromatin (*above*). *B*. M II, 23, XX. The haploid chromosome set of the secondary oocyte (see *A*.), arranged into a karyotype

*A*. × 1,300

The oocyte has now completed the first meiotic division, progressing from dictyotene through diakinesis, metaphase I, anaphase I, telophase I to metaphase II, where in vivo, there is a natural arrest up to the time of fertilization.

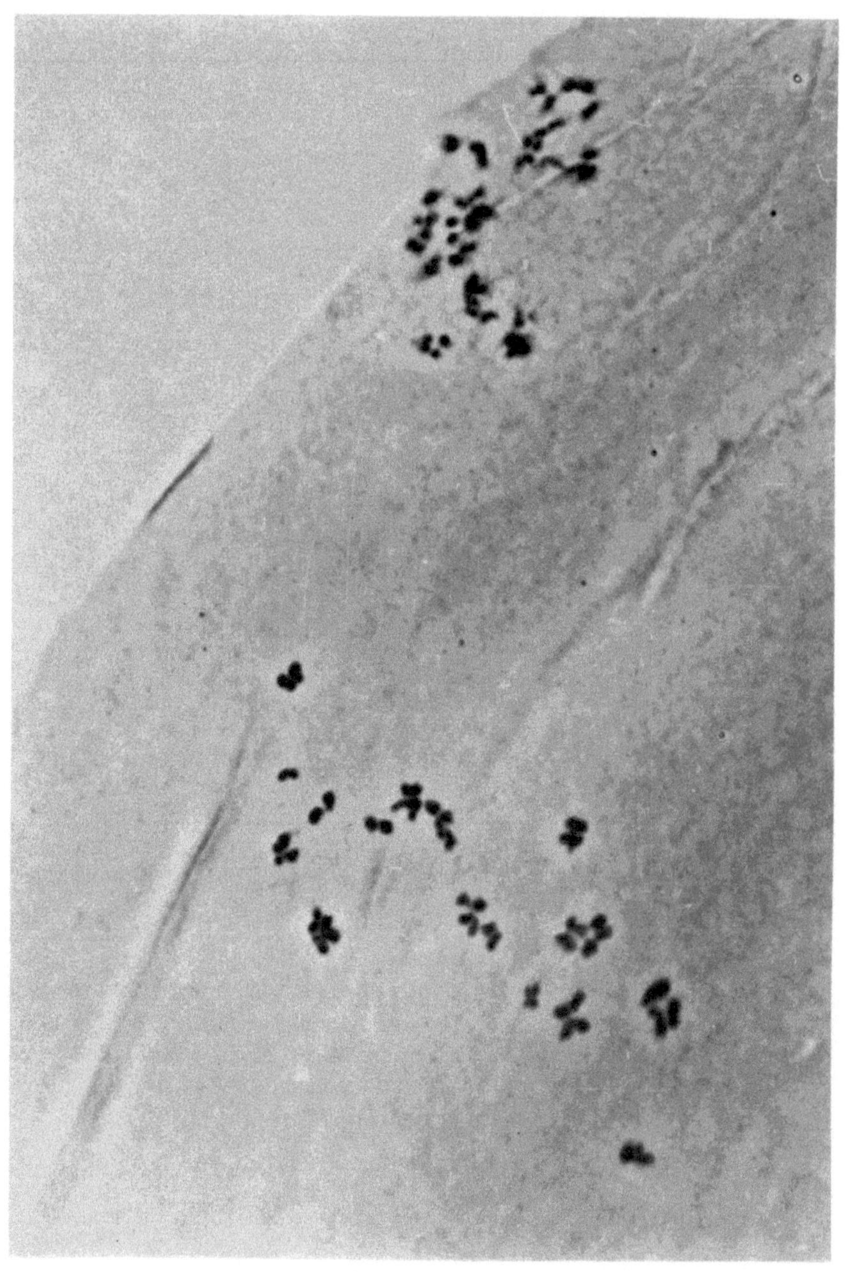

Fig. 51 A

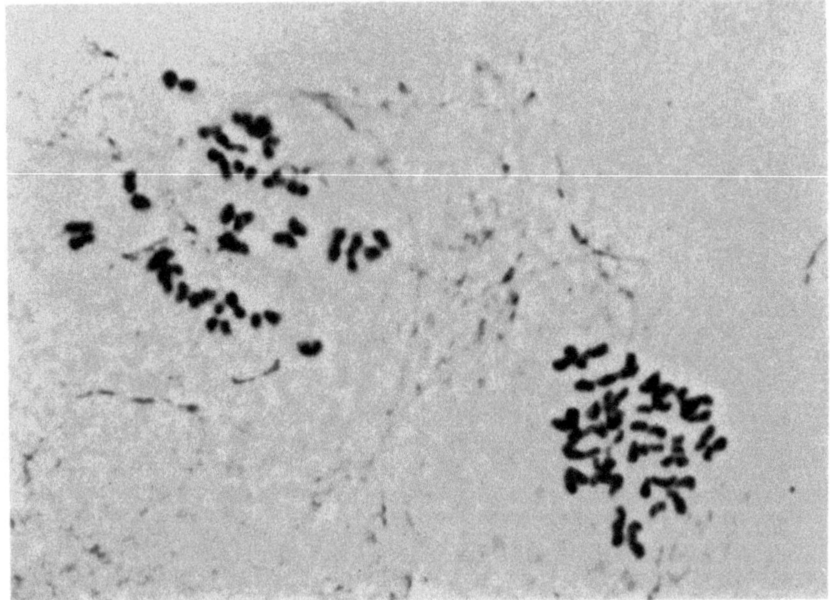

Fig. 52A. A mature oocyte at metaphase II. At the *right* are the chromosomes belonging to the secondary oocyte, at the *left* are the chromosomes of the first polar body. In this preparation the first polar body chromosomes, usually present as scattered, degenerating chromatin, were in a state that allowed karyotyping (see Fig. 52B)

× 1,500

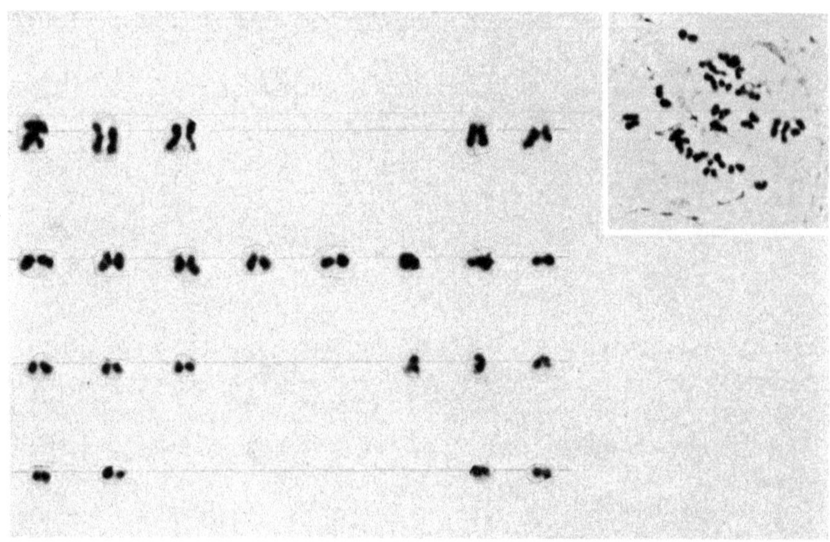



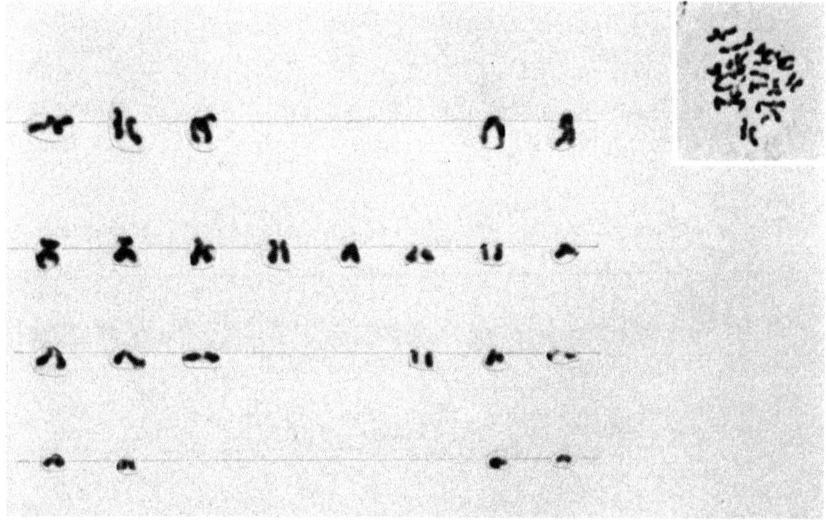



Fig. 52 B. First polar body chromosomes, arranged in a karyotype. Coiling and condensation are similar to those of the second metaphase (see Fig. 52 C)

Fig. 52 C. Metaphase II chromosomes, arranged in a karyotype. Regular distribution of the chromosomes to the oocyte and first polar body (Fig. 52 B) is evident. Some of the chromosomes have divided at the centromeres, indicating approaching anaphase of the second maturation division

Fig. 53. Metaphase II plate of a mature oocyte. The *arrow* indicates an association (*Ass.*) of acrocentric chromosomes. Three chromosomes participate: one of the D-group and two of the G-group; a small unidentified particle is lying between them. Their proximity in the metaphase plate may reflect the close relationship between nucleolus and chromosomes, as is observed throughout meiotic prophase (cf., among others, Figs. 15A, 16A, and 19)

× 3,900

The association of acrocentric chromosomes, already known from mitotic cells, has received much attention because of its apparent relation to chromosome anomalies

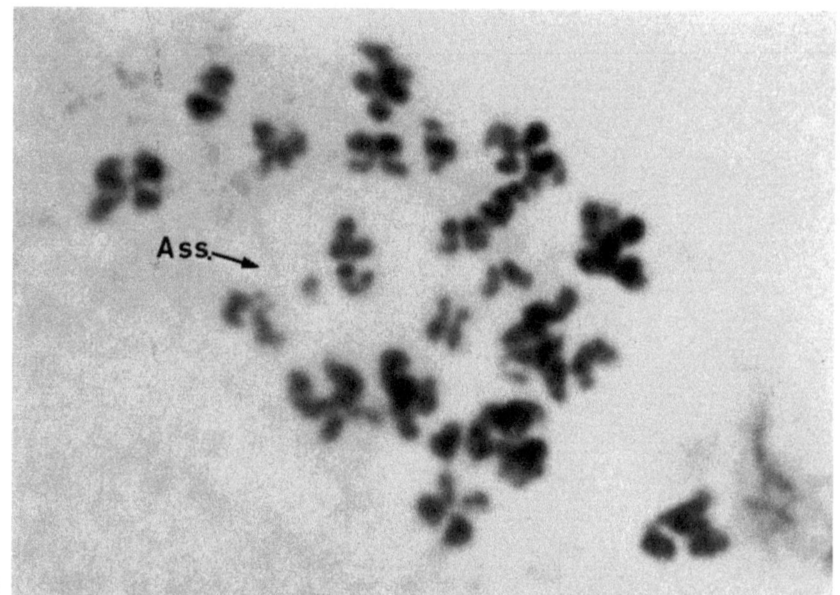



Fig. 53

# Selected References


Austin, C.R., Short, R.V. (eds.): Reproduction in Mammals. Book 1: Germ Cells and Fertilization. Cambridge: Cambridge Univ. Press 1972.

Biggers, J.D., Schuetz, A.W. (eds.): Oogenesis. Baltimore: University Park Press; London: Butterworths 1972.

Blandau, R.J. (ed.): Aging Gametes, Their Biology and Pathology. Basel-New York: Karger 1975.

Chandley, A.: Culture of mammalian oocytes. J. Reprod. Fert. (Suppl.) **14**, 1–6 (1971).

Channing, C.: Steroidogenesis and morphology of human ovarian cell types in tissue culture. J. Endocrinol. **45**, 297–308 (1969).

Donahue, R.P.: Reproductive Genetics. Birth Defects: Orig. Art. Ser., 1974, Vol. X, Nr. 10, pp. 3–13.

Eberle, P.: Die Chromosomenstruktur des Menschen in Mitosis und Meiosis. Fortschritte der Evolutionsforschung. Heberer, G. (ed.), Stuttgart: G. Fischer, 1966, Vol. II.

Edwards, R.G.: Observations on meiosis in normal males and females. In: Human Population Cytogenetics, Jacobs, P.A., Price, W.H., Law, P. (eds.), Edinburgh: Edinburgh University Press, 1970, Pfizer Medical Monographs **5**, pp. 9–21.

Edwards, R.G., Fowler, R.F.: The genetics of human preimplantation development. In: Modern Trends in Human Genetics, Emery, A.E.H. (ed.), London: Butterworths 1970, Vol. I, pp. 181–213.

Edwards, R.G.: Maturation in vitro of human ovarian oocytes. Lancet **1965 II**, 926–929.

Ferguson-Smith, M.A.: Chromosomal satellite association. Lancet **1967 I**/7500, 1156–1157.

Ford, E.H.R.: Human Chromosomes. London-New York: Academic Press 1973.

Fowler, R.E., Edwards, R.G.: The genetics of early human development. In: Progress in Medical Genetics, Steinberg, A.G., Bearn, A.G. (eds.), New York-London: Grune and Stratton, 1973, Vol. IX, pp. 49–112.

Gondos, B., Zamboni, L.: Ovarian development: The functional importance of germ cell interconnections. Fertil. Steril. **20**, 176–189 (1969).

Hamerton, J.L.: Human Cytogenetics. New York-London: Academic Press, 1971, Vols. I, II.

Hoehn, H., Sander, C., Sander, L.Z.: Aneusomie de recombinaison: Rearrangement between paternal chromosomes 4 and 18 yielding offspring with features of the 18q– syndrome. Ann. Genet. **14**, 187–192 (1971).

Hultén, M.: Cytogenetic aspects of human male meiosis. Thesis, Stockholm, 1974.

Jagiello, G., Ducayen, M., Fang, J.-S., Graffeo, J.: Cytogenetic observations in mammalian oocytes. In: Chromosomes Today V, Pearson, P.L., Lewis, K.L. (eds.), New York: Wiley, 1976, pp. 43–63.

John, B., Lewis, K.R.: The Meiotic System. Protoplasmatologia. Berlin-Heidelberg-New York: Springer, 1965, Vol. VI.

Kennedy, J.F., Donahue, R.P.: Human oocytes: Maturation in chemically defined media. Science **164**, 1292–1293 (1969).

Knörr, K., Knörr-Gärtner, H.: Das Abortgeschehen unter genetischen Aspekten. Gynäkologe **10**, 3–8 (1977).

Knörr-Gärtner, H., Knörr, K., Haas, B.: Familiäre Translokation t (4q – ; 18q +) mit verschiedenartigen unbalancierten Nachkommen. Humangenetik **21**, 315–321 (1974).

Knörr, K., Uebele-Kallhardt, B.: Meiosestörungen menschlicher Eizellen nach Einwirkung endogener und exogener Noxen. In: Fortschritte der Fertilitätsforschung, Schirren, C. (ed.), Berlin: Grosse, 1971, Vol. 2, pp. 28–33.

Knörr, K., Uebele-Kallhardt, B.: Entwicklung und Reifung der Eizelle und ihre Störungen. In: Funktion und Pathologie des Ovariums, König, P.A., Probst, V. (eds.), Stuttgart: Enke, 1971, pp. 6–12.

Langman, J.: Medical Embryology, 3rd ed., Baltimore: Williams and Wilkins.

Lejeune, J., Berger, R.: Sur deux observations familiales de translocations complexes. Ann. Génét. **8**, 21–29 (1965).

Nakogome, Y.: G-group chromosomes in satellite associations. Cytogenet. Cell Genet. **12**, 336 (1972).

Nakogome, Y.: Participation of D-group chromosomes in satellite associations. Humangenetik **25**, 235–236 (1974).

Nalbandov, A.V.: Interaction between oocytes and follicular cells. In: Oogenesis, Biggers, J.D., Schuetz, A.W. (eds.), Baltimore: University Park Press; London: Butterworths 1972, pp. 513–522.

Paris Conference (1971): Standardization in Human Cytogenetics. Birth Defects Orig. Art. Ser. New York: The National Foundation 1972, Vol. 8, Nr. 7.

Patil, S.R., Lubs, H.A.: Non-random association of human acrocentric chromosomes. Humangenetik **13**, 157–159 (1971).

Pincus, G., Saunders, B.: The comparative behaviour of mammalian eggs in vivo and in vitro. II. The maturation of human ovarian ova. Anat. Rec. **75**, 537–545 (1939).

Pinkerton, H.M., McKay, D.G., Adams, E.C., Hertig, A.T.: Development of the human ovary—a study using histochemical techniques. Obstet. Gynec. **18**, 152–181 (1961).

Robinson, J.A.: Meiosis I non-disjunction as the main cause of trisomy 21. Hum. Genet. **39**, 27–30 (1977).

Sasaki, M., Makino, S.: The meiotic chromosomes of man. Chromosoma **16**, 637–651 (1965).

Sele, B., Jalbert, P., van Cutsem, B., Lucas, M., Mouriquand, C., Bouchez, R.: Distribution of human chromosomes on the metaphase plate using banding techniques. Hum. Genet. **39**, 39–63 (1977).

Shettles, L.B.: Ovum Humanum. München-Berlin: Urban und Schwarzenberg 1960.

Suzuki, S.: An Atlas of Mammalian Ova. Stuttgart: Thieme 1974.

Tsafiri, A., Pomerantz, S.H., Channing, C.P.: Follicular control of oocyte maturation. In: Ovulation in the Human, Crosignani, P.G., Mishell, D.R. (eds.), London-New York: Academic Press, 1976, pp. 31–39.

Uebele-Kallhardt, B., Knörr, K.: Meiotische Chromosomen der Frau. Humangenetik **12**, 182–187 (1971).

Uebele-Kallhardt, B.: Mutation frequency in human oocytes. Humangenetik **16**, 127–129 (1972).

Uebele-Kallhardt, B., Knörr, K.: Meiotic chromosome study in a human female translocation heterozygote. Humangenetik **26**, 355–356 (1975a).

Uebele-Kallhardt, B., Knörr, K.: Die Eizellen polycystischer menschlicher Ovarien. Arch. Gynäk. **218**, 189–201 (1975b).

Wieczorek, V.: Chromosomenanomalien als Ursache von Fehlgeburten. München: Goldmann 1971.

Yuncken, C.: Meiosis in the human female. Cytogenetics **7**, 234–238 (1968).

Zamboni, L.: Fine structure of human follicular oocytes maturing in vitro. In: Biology of Reproduction, Biggers, J.D.(ed.), New York-London: Academic Press, 1971, Vol. 5, p. 90.

Zamboni, L.: Modulations of follicle cell-oocyte association in sequential stages of mammalian follicle development and maturation. In: Ovulation in the Human, Crosignani, P.G., Mishell, D.R. (eds.), London-New York: Academic Press, 1976, pp. 1–30.

# Materials and Methods

The present study of the *first meiotic prophase* has covered a period of several years, from 1971 to 1975. The ovaries were obtained from spontaneous or therapeutic abortions of human fetuses aged from 18 to 27 weeks.

Immediately after removal the ovaries were placed in an isotonic salt solution and minced until a fine suspension was produced. After separating the larger fragments nuclear preparations were made by the usual suspension technique, avoiding any hypotonic treatment to preserve the fragile chromosome structures and the nucleolar–chromosomal relationships. Air-dried preparations were stained with aceto-orcein.

With this procedure, the slides generally contain numerous oocytes in a variety of meiotic stages.

The stages of the *first and second meiotic divisions* were selected from more than 2,000 human oocytes, cultured in vitro. Either the whole ovary or small biopsies were obtained from patients in the 20-to-45 year age range, undergoing gynecologic surgery. The ovary specimens were rinsed in saline solution, the oocytes removed from their follicles, and examined under a dissecting microscope. Grossly degenerate oocytes were not cultured. Oocytes with a normal appearance were transferred into drops of a suitable medium, such as Eagle's MEM completed with serum, which had previously been placed under paraffin oil in tissue culture dishes (Figs. 54, 55). The microcultures were incubated in a gas phase of 5% $CO_2$ in air at 37° C. Metaphase I generally was reached at 25–27 h, metaphase II at about 40 h of culture. According to the desired nuclear stage, the oocytes were taken from the culture, put into hypotonic solution and then fixed on a slide. Air-dried preparations were stained in aceto-orcein.

The staining techniques for obtaining chromosome banding, applied to the maximally spiralized metaphase chromosomes usually present in cultured human oocytes, thus far has yielded unsatisfactory results.

Photographic documentation: Living follicular oocytes were photographed before and after culture on Kodak Plus-X Pan film with a Zeiss inverse microscope. Photographs from chromosome preparations of both meiotic prophase and maturation divisions were taken on Copex Pan film with a Zeiss photomicroscope, either with bright field or with phasecontrast, depending on the different degree of contraction and the stainability of the chromosomes.

Fig. 54. Transfer of oocytes into a drop of culture medium under paraffin oil

Fig. 55. Examination of oocyte microcultures by an inverse microscope

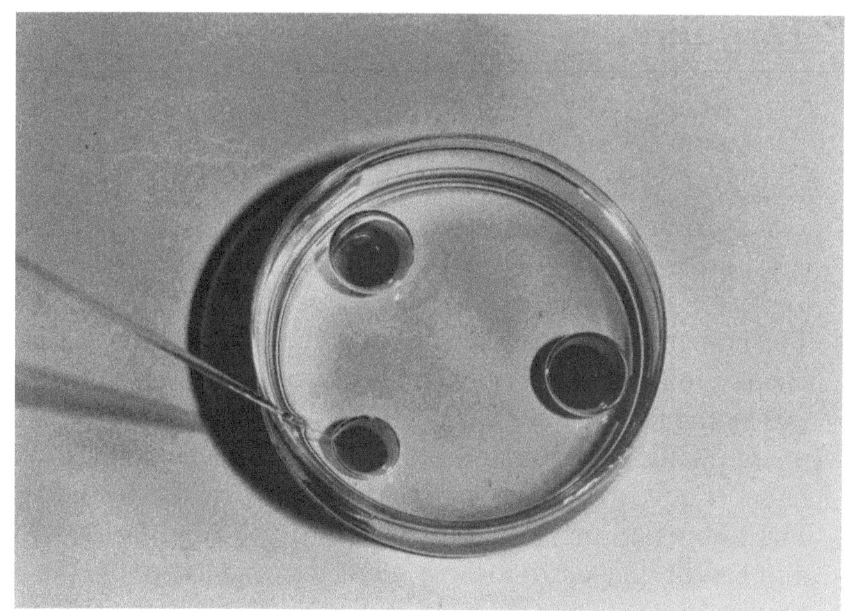

Fig. 54

Fig. 55

# Terminology

The nomenclature used in the present study of human female meiosis follows the fundamentals given for male meiotic analysis by the Paris Conference, 1971: M I indicates the first metaphase, including diakinesis, M II the second metaphase. PB I refers to the first polar body, extruded from oocytes in M II. The abbreviations M I and M II are followed by the total number of the chromosome elements and then by the sex chromosomes. Thus, M I, 23, XX indicates a primary oocyte at diakinesis or first metaphase with 23 elements including the XX bivalent. The latter cannot be distinguished from the autosomal bivalents by conventional staining techniques.

To describe abnormal cells, the additional abbreviations I, II, III, IV are used, which denote univalents, bivalents, trivalents, and quadrivalents, respectively. The numerals are followed in parentheses by the chromosomes involved in a rearrangement. Thus, M I, 22, IV (4; 18) indicates a primary oocyte at first metaphase with 22 elements including the XX bivalent and a quadrivalent, consisting of the chromosomes 4 and 18.

# Selected References

Luciani, J.M., Devictor-Vuillet, M., Gagné, R., Stahl, A.: An air-drying method for first meiotic prophase preparations from mammalian ovaries. J. Reprod. Fertil. **36**, 409–411 (1974).
Williams, D.L., Lafferty, D.A., Webb, S.L.: An air drying method for the preparation of dictyotene chromosomes from ovaries of chinese hamsters. Stain Technol. **45**/3, 133–135 (1970).

Edwards, R.G.: Maturation in vitro of human ovarian oocytes. Lancet **1965 II**, 926–929.
Jagiello, G., Karnicki, J., Ryan, R.: Superovulation with pituitary gonadotrophins. Method for obtaining metaphase figures in human ova. Lancet **1968 I**, 178–180.
Steptoe, P.C., Edwards, R.G.: Laparoscopic recovery of preovulatory human oocytes after priming of ovaries with gonadotrophins. Lancet **1970 I**, 683–690.
Tarkowsky, A.K.: An air-drying method for chromosome preparations from mouse eggs. Cytogenetics **5**, 394–400 (1966).

# Subject Index

Aberration  1, 92
— nucleolar-chromosomal  28–29
— numerical  42
— structural  42, 74–77
Abnormalities chromosomal, see Aberration
Acrocentric chromosomes  3, 24–29, 32–33
— — association of  92–93
— — and nucleoli  24–29, 32–33
Anaphase I, disjunction  32, 41, 80, 82–83
— — non-disjunction  42
„Aneusomie de Recombinaison" 76
Arrest, meiotic  1, 4, 34, 44–45, 56–57, 88
Atresia, follicular  46, 48, 50

Banding pattern, see Technique
Bivalent  3, 4, 22, 66–69, 72–73
— acrocentric  3, 24–29, 32–33
Bouquet stage  20–21
Breakage chromosomal, meiotic consequences  76–77

Carrier, see Translocation
Chiasma  1, 3, 30, 32–35, 68–69
— at diakinesis  68–69
— at diplotene  32–35
— frequency in human oocytes 42
— terminalization  78–79

Chromomere  30–31
— terminal  32–33
Chromosomes, see under required heading
Coiling cycle, meiotic  1–2
— — despiralization phases 16–19, 32–35
— — spiralization phases  10–15, 20–22, 64–67, 73, 79, 89
Corona radiata  44–45
— — loss of 48–49, 52–53
Crossing over, see Chiasma
Culture of oocyte  58–59, 62–63, 99–101
— of follicle cells  60–63
Cumulus oophorus  44–47
Cytoplasm, division  1, 85
— fragmentation  52–55
— shrinkage  48–53

Diakinesis  41, 68–69
Dictyotene  1, 4, 34, 41, 44–45, 56–57, 64
— arrest meiotic  1, 4, 44–45, 56–57
— nucleus, see Germinal vesicle
— structure of chromosomes  4, 34
Diplotene  3, 4, 30, 32–35
— nucleoli  32–35
— — rDNA  34
Disjunction, see Anaphase I
DNA-replication  3
Dyad  41

Follicle   43, 44, 56, 62, 99
— atretic   46, 48, 50
— biovular   46–47
— primary   4
Follicle cell   44–45, 50–51, 54, 58
— — culture   60–63
— — necrosis   60
— — nutritive function   54
— — process   54–55
Fragmentation, see Cytoplasm

Gamete, balanced   76–77
— unbalanced   76–77
Gene action, see Chromomere
Germ cell   3, 4, 5
— — degeneration   41
— — population   41, 42
— —primordial   6
Germinal vesicle   34, 56–57, 64–65

Interphase premeiotic   3

Karyotype, first metaphase   72
— — — translocation carrier   75
— second metaphase   88, 91
— first polar body   91

Lampbrush chromosomes   4, 34
Leptotene   3, 8–9, 10, 18–19

Meiosis, arrest   1, 4, 34, 44–45,
   56–57, 88
— duration in oocyte   1
— first division   1, 41–42, 64–83
— preovulatory stages   2, 41
— prophase   1, 2, 6, 8–35, 64–69
— resumption   1, 41, 56, 64
— second division   1, 2, 41–42,
   84–93
Metaphase, first division   72–75,
   78–79
— second division   88–93
Mutation, see Translocation
— frequency in oocytes   42

Necrosis, see Follicle cell
Nomenclature for meiotic chromosome complement   102
Non-disjunction, see Anaphase I
Nuclear membrane   41
Nucleolar-chromosomal relationship   3, 12–15, 24–29, 32–33,
   92–93
Nucleolus   12–15, 24–29, 41,
   64–67
— number in meiotic prophase   4,
   12
— organizer region   24
— primary (or main-)   32–35
— RNA synthesis   12
— secondary (or micro-)   34–35
Nucleus   1
— at start of culture   64–67
— dictyotene, see Germinal vesicle
— oogonium   6–7
— primary oocyte   6, 8–35,
   56–57, 64–67

Oocyte, degeneration   41, 48–55,
   62–63, 80–81
— duration of meiosis   1
— maturation in vitro, see Technique
— metabolic and synthetic activity   4
— number   41, 42
— primary   1, 4, 6, 8–9, 58–59
— secondary   41, 84–85, 88–91
— size   4
Oogonium   6–7
— chromosome number   6
— proliferation and transformation   3, 6
Ooplasm, see Cytoplasm

Pachytene   3, 8, 20, 22–31, 32
Pairing of chromosomes, see Synapsis

Photographic documentation, see
    Technique
Polar body, first  41, 42, 82, 84–91
— — — Karyotype  90–91
Polarization of chromosomes
    20–21
Polycystic ovaries (Stein-Leven-
    thal-Syndrome),
— — biovular follicle  46–47
Preleptotene  3, 8, 10–19
Preparation of meiotic chromoso-
    mes, see Technique
Prochromosomes  3, 8, 14–17
Prophase  3, 4, 6, 8–35, 68–69
— chromosome spiralization, see
    Coiling cycle
— stages: Diakinesis  41, 68–69
— — Dictyotene  1, 4, 34, 41,
    44–45, 56–57, 64
— — Diplotene  3, 4, 30, 32–35
— — Leptotene  3, 8–10, 18–19
— — Pachytene  3, 8–9, 20,
    22–31, 32
— — Preleptotene  3, 8–9, 10–19
— — Zygotene  3, 20–21

Quadrivalent, in complex translo-
    cation  42, 74–77

Recombination  76
Reduction division  82
rDNA (ribosomal DNA), see Di-
    plotene
RNA (ribonucleic acid), see Nucle-
    olus

Spindle formation  41
Synapsis, meiotic  3, 20, 22, 76

Staining of chromosomes, see
    Technique
Stickiness, chromosomal  80–81

Technique for chromosome study,
    first and second divisions
    86–87, 99–101
— — — — prophase  99
— oocyte culture  99–101
— photographic documentation
    100
— references  103
— staining for Banding pattern
    100
— timing of oocyte maturation in
    vitro  99
Telophase I  41
Terminalization, see Chiasma
Tetrad  22, 68–69
Timing of oocyte maturation in vi-
    tro  42, 99
Transcription, genetic, see Coiling
    cycle
Translocation  42, 74–77
— carrier, balanced  74
— formation of quadrivalent
    74–77
— heterozygosity  1
— meiotic consequences  74–77

Univalent  42

Zona pellucida  50–51, 54, 70–71,
    84–85
Zygotene  3, 20–21

# MGG

## *Molecular & General Genetics*

An International Journal

Continuation of Zeitschrift für Vererbungslehre
The first Journal on Genetics Founded in 1908

ISSN 0026-8952        Title No. 438

*Editorial Board:* W. Arber, Ch. Auerbach, E. Bautz,
H. Böhme, H. W. Boyer, B. A. Bridges, A. J. Clark,
W. Gajewski, W. Gehring, W. Gilbert, D. Goldfarb,
M. M. Green, F. Gros, K. Illmensee, F. Kaudewitz,
L. S. Lerman, G. Melchers, G. A. O'Donovan,
O. Siddiqi, F. W. Stahl, H. Stube, M. Weiss,
H. G. Wittmann, T. Yura

*Managing Editors:* G. Melchers, Tübingen and
H. Böhme, Gatersleben.

*Editorial Assistant:* H. Atzler, Tübingen

*Molekular and General Genetics* is a continuation of
*Zeitschrift für induktive Abstammungs- und Verbungs-
lehre,* the first international journal on genetics. The
current title reflects the role of molecular genetics and
the ever-growing emphasis on biophysical and biochemical aspects in the study of modern genetics. As
this journal also publishes original contributions
dealing with molecular aspects in genetics of higher
organisms, it serves as a handy source for keeping
abreast of all significant developments in this important
field.

*Fields of Interest:* General Genetics, Molecular
Genetics, Molecular Biology, Biochemistry, Bio-
physics, Developmental Physiology, Virology, Micro-
biology, Botany, Zoology.

## Springer-Verlag
## Berlin
## Heidelberg
## New York

*Subscription Information:*
1978. Vols. 158-165 (3 issues each):
DM 1504,–, plus postage and handling.
North America:
US $ 683.00, including postage and handling.